Vol. 29. **The Analytical Chemistry of Sulfur and Its Compounds** (*in three parts*). By J. H. Karchmer

Vol. 30. **Ultramicro Elemental Analysis.** By Günther Tölg

Vol. 31. **Photometric Organic Analysis** (*in two parts*). By Eugene Sawicki

Vol. 32. **Determination of Organic Compounds: Methods and Procedures.** By Frederick T. Weiss

Vol. 33. **Masking and Demasking of Chemical Reactions.** By D. D. Perrin

Vol. 34. **Neutron Activation Analysis.** By D. De Soete, R. Gijbels, and J. Hoste

Vol. 35. **Laser Raman Spectroscopy.** By Marvin C. Tobin

Vol. 36. **Emission Spectrochemical Analysis.** By Morris Slavin

Vol. 37. **Analytical Chemistry of Phosphorus Compounds.** Edited by M. Halmann

Vol. 38. **Luminescence Spectrometry in Analytical Chemistry.** By J. D. Winefordner, S. G. Schulman and T. C. O'Haver

Vol. 39. **Activation Analysis with Neutron Generators.** By Sam S. Nargolwalla and Edwin P. Przybylowicz

Vol. 40. **Determination of Gaseous Elements in Metals.** Edited by Lynn L. Lewis, Laben M. Melnick, and Ben D. Holt

Vol. 41. **Analysis of Silicones.** Edited by A. Lee Smith

Vol. 42. **Foundations of Ultracentrifugal Analysis.** By H. Fujita

Vol. 43. **Chemical Infrared Fourier Transform Spectroscopy.** By Peter R. Griffiths

Vol. 44. **Microscale Manipulations in Chemistry.** By T. S. Ma and V. Horak

Vol. 45. **Thermometric Titrations.** By J. Barthel

Vol. 46. **Trace Analysis: Spectroscopic Methods for Elements.** Edited by J. D. Winefordner

Vol. 47. **Contamination Control in Trace Element Analysis.** By Morris Zief and James W. Mitchell

Vol. 48. **Analytical Applications of NMR.** By D. E. Leyden and R. H. Cox

Vol. 49. **Measurement of Dissolved Oxygen.** By Michael L. Hitchman

Vol. 50. **Analytical Laser Spectroscopy.** Edited by Nicolo Omenetto

Vol. 51. **Trace Element Analysis of Geological Materials.** By Roger D. Reeves and Robert R. Brooks

Vol. 52. **Chemical Analysis by Microwave Rotational Spectroscopy.** By Ravi Varma and Lawrence W. Hrubesh

Vol. 53. **Information Theory As Applied to Chemical Analysis.** By Karel Eckschlager and Vladimir Štěpánek

Vol. 54. **Applied Infrared Spectroscopy: Fundamentals, Techniques, and Analytical Problem-solving.** By A. Lee Smith

Vol. 55. **Archaeological Chemistry.** By Zvi Goffer

Vol. 56. **Immobilized Enzymes in Analytical and Clinical Chemistry.** By P. W. Carr and L. D. Bowers

Vol. 57. **Photoacoustics and Photoacoustic Spectroscopy.** By Allan Rosencwaig

Vol. 58. **Analysis of Pesticide Residues.** Edited by H. Anson Moye

Vol. 59. **Affinity Chromatography.** By William H. Scouten

Vol. 60. **Quality Control in Analytical Chemistry.** By G. Kateman and F. W. Pijpers

Vol. 61. **Direct Characterization of Fineparticles.** By Brian H. Kaye

Vol. 62. **Flow Injection Analysis.** By J. Ruzicka and E. H. Hansen

(*continued on back*)

Remote Sensing by
Fourier Transform Spectrometry

CHEMICAL ANALYSIS

A SERIES OF MONOGRAPHS ON
ANALYTICAL CHEMISTRY AND ITS APPLICATIONS

Editor
J. D. WINEFORDNER
Editor Emeritus: **I. M. KOLTHOFF**

VOLUME 120

A WILEY-INTERSCIENCE PUBLICATION

JOHN WILEY & SONS, INC.

New York / Chichester / Brisbane / Toronto / Singapore

Remote Sensing by Fourier Transform Spectrometry

REINHARD BEER

Earth and Space Sciences Division
Jet Propulsion Laboratory
California Institute of Technology
Pasadena, California

A WILEY-INTERSCIENCE PUBLICATION

JOHN WILEY & SONS, INC.

New York / Chichester / Brisbane / Toronto / Singapore

Library of Congress Cataloging in Publication Data:

Beer, Reinhard, 1935–
    Remote sensing by Fourier transform spectrometry / Reinhard Beer.
        p.   cm. — (Chemical analysis, ISSN 0069-2883 ; v. 120)
    "A Wiley-Interscience publication."
    Includes bibliographical references and index.
    ISBN 0-471-55346-8
    1. Fourier transform spectroscopy. 2. Remote sensing. I. Title.
II. Series.
QD96.F68B43   1991   91-33607
621.36′78—dc20   CIP

Printed and bound in the United States of America by Braun-Brumfield, Inc.

10 9 8 7 6 5 4 3 2 1

To the memory of Bob Norton:
my friend, collaborator, and mentor for nearly 30 years

FOREWORD

Fourier transform spectrometry (FTS) is accepted nowadays by spectrometrists in many fields as a reliable technique with powerful advantages. This acceptance is a distinct change from the scepticism with which FTS was treated in its early years of application by astronomers and atmospheric physicists who had been brought up on grating instruments. It was a common reaction that you couldn't "see" the spectrum and that one had no way of knowing whether the data that had been recorded were good or bad until they had been put through an awkward process of mathematical transformation. The development of sophisticated electro-optical control systems and, in particular, of computers capable of handling the transformation of interferograms of up to several million points in quasi-real time has brought the FTS technique to the state where, today, it is the standard method of spectrometry at almost all frequencies throughout the electromagnetic spectrum.

The application of the FT method has played a very significant role in the advances that have been made in the study of planetary atmospheres, including our own. Among these are the discovery of HCl and HF on Venus by ground-based observations, the detection of the same constituents in the Earth's stratosphere some 10 years later by aircraft and balloon-borne instruments, heavy hydrocarbons in the atmospheres of the outer planets by the *Voyager* spacecraft, and the beautiful extended bands of CO seen in the spectrum of the sun recorded from above the Earth's atmosphere by the ATMOS interferometer carried by the shuttle *Challenger*. These are but a few examples of the application of FTS to remote sensing in situations where the constraints on the observation—available time and incident energy—render the characteristic advantages of the FT method indispensable.

While there have been many publications dealing with the instrumental and mathematical processing aspects of FTS, there remains a need for a text that brings the salient points of the technique as a remote sensing tool to students and others in such fields as geophysics and astrophysics who are considering the application of FTS to their research needs. In particular, this book provides—in one place—a concise summary of the fundamentals of the theoretical basis, optical design, sampling, and data processing aspects of FTS. Reinhard Beer's long experience in the development of FT instruments and their application to problems in planetary and Earth atmosphere research has given him a unique viewpoint from which his series of JPL lectures on the subject could be presented. It was the success of these that stimulated the suggestion that he publish his lecture notes as a book for wider dissemination. As such, it will be especially useful as a starting point for those coming into the field of experimental high-resolution spectrometry of atmospheres.

CROFTON B. FARMER

Glendale, California
January 1992

PREFACE

This book had its genesis as a set of notes for a five-part course taught at the Jet Propulsion Laboratory (JPL) in the fall of 1989; the notes have been expanded and updated for this book. The original motivation was that, while JPL has been a leader in the development and utilization of Fourier transform spectrometers (FTS) for remote sensing of planetary atmospheres and surfaces, nevertheless knowledge of their properties is still shared by relatively few people. However, the advent of new requirements stemming from the well-known set of problems commonly called *global change* has resulted in new generations of FTS being proposed and accepted for a wide variety of remote sensing platforms (primarily balloons, aircraft and spacecraft). These new requirements have, in turn, engendered a wide range of new technologies, many of them unique to the field. Since I am one of the generators of these new requirements and have spent most of my career both building and using such systems in a variety of environments, I felt it to be both timely and appropriate to attempt to provide an overview of the field with particular emphasis on the technology of optimum data acquisition. The related field of data processing and geophysical parameter retrieval is discussed only briefly since it is a discipline unto itself. Also neglected is the far larger field of laboratory and *in situ* usage of FTS.

While I have tried to minimize the number of acronyms the reader must contend with, a few are useful. In particular, as already noted, the term Fourier transform spectrometer will be abbreviated throughout as FTS. It is a personal idiosyncracy that I dislike the acronym FTIR (Fourier transform infrared, much used by chemical spectrometrists) because the field now encompasses the ultraviolet to the microwave region. I also prefer the term *spectrometry* to *spectroscopy*: modern techniques *measure* the spectra; visual examination (while great fun) is becoming rare.

Chapter 1 begins at a very elementary level, primarily to introduce the reader to the terminology used throughout the book, including a brief digression on the units common to the field; and ends with introductions to the Michelson interferometer as the archetypical FTS and both integral and discrete Fourier transforms.

Chapter 2 builds on this base by comparing the properties of an idealized FTS with equally ideal dispersive (grating) spectrometers. Introduced here are the concepts of étendue, multiplexing, spectral resolution, modulation efficiency, and implementation of imaging spectrometers.

Chapter 3 is a major digression into the basics of atmospheric physics and chemistry, because it is important to understand the origins of the spectra to be measured. The emphasis is strongly on atmospheres because it is in the field of atmospheric remote sensing that the FTS excels and, in any case, it is the field of my personal interest.

Chapter 4 covers some of the many problems that can arise in the design, development, and implementation of FTS. Beginning with an unconventional approach to the determination of signal-to-noise ratio, the chapter continues with a discussion of the all-important topic of interferogram sampling, followed by a discussion of some optical configurations widely used in remote sensing FTS systems. The chapter concludes with a look at some technical issues that have bedeviled many an FTS (including some of my own!).

Chapter 5 provides an overview of four existing and one developmental FTS as examples of how some investigators have addressed the requirements for some very different uses and operating environments. While there are certainly many more I could have chosen to discuss, these five are systems with which I have some personal involvement or familiarity.

Chapter 6 discusses the environments in which remote sensors must operate. While not entering into details of mounting, accommodations, and logistics, it does offer some personal experiences that may help potential users to avoid some of my own errors of omission and commission.

Finally, Chapter 7 provides a brief conclusion, followed by an appendix discussing the optimum filter theorem and its pertinence to spectrometry. It is, I believe, the only valid approach to the often-asked question, "What spectral resolution do I need to address my problem?"

I have therefore aimed the text more at optical and instrumentation scientists and engineers rather than remote sensing scientists per se. Nevertheless, it is my hope that students of the field and anyone considering the applicability of this methodology to his or her own field will obtain some benefit from reading this book.

Fourier transform spectrometry has many subtleties that cannot be covered in such an introductory text. Readers should not, therefore, expect to learn enough to design their own systems simply by reading this book alone, but they will be aware of the many pitfalls by which the unwary may be caught.

REINHARD BEER

Pasadena, California
March 1992

ACKNOWLEDGMENTS

This book was written with the support and encouragement of the Jet Propulsion Laboratory, California Institute of Technology, to which I offer my thanks for years of unstinting support.

In addition, I would be remiss were I not to acknowledge the help I have received over the past quarter-century from many individuals. While they are too many to list, I wish to offer my personal tribute to the late Robert H. Norton, who analyzed everything: from our first stumbling efforts in this field; through our years in astronomy together; and on to the more recent era of Earth observation. I must also thank C. B. Farmer of JPL, who introduced me to the field of Earth atmosphere analysis and contributed the Foreword to this book; to David G. Murcray of the University of Denver, who has been a generous friend and selfless collaborator; to Rudi Hanel for correcting some errors in Chapter 5; and, above all, to Pierre and Janine Connes of CNRS, France, who taught all of us how it should be done.

CONTENTS

CHAPTER 1 THE BASIC PRINCIPLES OF FOURIER TRANSFORM SPECTROMETRY   1

  1.1. Electromagnetic Waves   1

  1.2. The Principle of Superposition   2

  1.3. Units   3

    1.3.1. Spectrometric Units   4

    1.3.2. The Units of Atmospheric Chemistry and
        Physics   4

  1.4. Two-Beam Interference   5

  1.5. The Michelson Interferometer   6

    1.5.1. Basic Implementation as a Spectrometer   7

    1.5.2. Finite Field of View   8

    1.5.3. Finite Spectral Bandwidth   10

  1.6. The Integral Fourier Transform   10

  1.7. Convolution   12

    1.7.1. Physical Interpretation   13

  1.8. The Discrete Fourier Transform   14

CHAPTER 2 THE IDEAL FOURIER TRANSFORM SPECTROMETER   15

  2.1. Étendue   15

    2.1.1. Beam Area   16

    2.1.2. Solid Angle   17

    2.1.3. Comparison to Dispersive Systems   18

  2.2. Multiplexing   19

    2.2.1. Implementation   19

2.2.2. A Warning 20

2.2.3. Terminology 20

2.3. Spectral Resolution and Instrumental Line Shape 20

2.3.1. Spectral Resolution 20

2.3.2. Instrumental Line Shape 22

2.3.3. Consequences 22

2.4. Frequency Accuracy 22

2.4.1. Fourier Transform Spectrometers 23

2.4.2. Dispersive Spectrometers 23

2.5. Modulation Efficiency 24

2.5.1. The Impact of Impaired Modulation 25

2.6. Imaging Modes 25

2.6.1. FTS Implementation 25

2.6.2. Another Warning 28

2.6.3. Dispersive Spectrometer Implementation 28

2.6.4. Discussion 29

CHAPTER 3 THE PHYSICS AND CHEMISTRY OF REMOTE SENSING 31

3.1. The Physics of Atmospheres 31

3.1.1. The Atmospheric Temperature Profile 31

3.1.2. The Atmospheric Pressure and Density Profile 33

3.1.3. Model Atmospheres 34

3.2. The Chemistry of Atmospheres 34

3.2.1. Atmospheric Composition 35

3.2.2. Compositional Profiles 35

3.3. Radiative Transfer 36

3.3.1. Radiative Transfer in a Clear Atmosphere 39

3.3.2. Discussion 42

3.4. Remote Sensing Spectrometry 44

3.4.1. The Mechanistic Molecular Model 44

3.4.2. Intensities 46

3.4.3. Terminology 47

3.4.4. Line Shapes   47
3.4.5. Quantitative Atmospheric Spectral
        Analysis   51
3.4.6. The Spectra of Solids and Liquids   52
3.4.7. Discussion   54

CHAPTER 4   **REAL FOURIER TRANSFORM
            SPECTROMETERS**   **55**

4.1. The Estimation of Signal-to-Noise Ratio   55
      4.1.1. Sources and Backgrounds   55
      4.1.2. Detectors   58
      4.1.3. Radiometric Models   60
      4.1.4. Fourier Transform Spectrometer   62
      4.1.5. Dispersive Spectrometer   66
      4.1.6. The Calculation of $D*$ and Background
              Flux Density   67
      4.1.7. The Impact of Pointing Jitter   68
      4.1.8. Discussion   69
4.2. Interferogram Sampling   69
      4.2.1. Sampling Theory   70
      4.2.2. A Brief Digression   74
      4.2.3. Sampling Position Accuracy   75
      4.2.4. Sampling Strategy   76
      4.2.5. Sampling Methodology   77
      4.2.6. Symmetric and Asymmetric Scanning   82
4.3. Optical Configurations   83
      4.3.1. The Michelson FTS   83
      4.3.2. Retroreflectors   83
      4.3.3. The Connes-Type FTS   85
      4.3.4. The Compensated FTS   87
      4.3.5. Discussion   89
4.4. Potential Problem Areas   89
      4.4.1. Phase Errors   89
      4.4.2. Channeling   91

4.4.3. Vibration   91
4.4.4. Electromagnetic Interference   93
4.4.5. Signal Dynamic Range   94
4.4.6. Signal Chain Linearity   97
4.4.7. Data Errors   98
4.4.8. Data Rates and Volumes   99

CHAPTER 5   CASE STUDIES OF REMOTE SENSING
             FOURIER TRANSFORM SPECTROMETERS   101

5.1. The *Voyager* IRIS FTS   101
      5.1.1. Purpose   101
      5.1.2. Specifications   102
      5.1.3. Optical Layout   102
      5.1.4. Discussion   104
5.2. The Canada–France–Hawaii Telescope FTS   104
      5.2.1. Purpose   107
      5.2.2. Specifications   107
      5.2.3. Optical Layout   107
      5.2.4. Discussion   108
5.3. The *Spacelab 3* ATMOS FTS   109
      5.3.1. Purpose   113
      5.3.2. Specifications   113
      5.3.3. Optical Layout   114
      5.3.4. Discussion   114
5.4. The Mark IV Balloon/Aircraft FTS   114
      5.4.1. Purpose   116
      5.4.2. Specifications   117
      5.4.3. Optical Layout   117
      5.4.4. Discussion   117
5.5. The *EOS* TES FTS   121
      5.5.1. Purpose   123
      5.5.2. Specifications   124
      5.5.3. Optical Layout   124
      5.5.4. Discussion   124

CHAPTER 6 REMOTE SENSING ENVIRONMENTS 129

    6.1. Telescope-Based Systems 129
    6.2. Aircraft-Based Systems 131
    6.3. Balloon-Borne Systems 132
    6.4. Spacecraft Systems 133

CHAPTER 7 GENERAL OBSERVATIONS AND CONCLUSIONS 135

APPENDIX OPTIMUM FILTERS 137

    A.1. The Optimum Filter Theorem 137
    A.2. Implementation 140
    A.3. An Alternative Approach 142

BIBLIOGRAPHY AND REFERENCES 145

INDEX 149

Remote Sensing by
Fourier Transform Spectrometry

CHAPTER

1

THE BASIC PRINCIPLES OF FOURIER TRANSFORM SPECTROMETRY

This opening chapter outlines some of the basic physical and mathematical principles upon which Fourier transform spectrometers rely. While you may be tempted to skip this chapter because it begins at a very elementary level, I nevertheless suggest you read it because it is here that much of the terminology and units used throughout the book are introduced.

1.1. ELECTROMAGNETIC WAVES

The propagation of electromagnetic waves is governed by Maxwell's equations. We shall not discuss the equations themselves, only their utilization.

The most common solution of Maxwell's equations is that of a sinusoidal wave propagating through free space (i.e., a perfect vacuum) at a velocity $c = 2.997925 \times 10^8 \text{ m·s}^{-1}$. Associated with this wave is a *frequency* v Hz (1 Hz $\equiv$ 1 cycle per second) and a *wavelength* λ meters such that

$$\lambda = c/v \qquad (1.1)$$

If the wave travels through any other medium, the velocity is reduced to a value c'. Since the frequency is deemed to be invariant (i.e., a property of the source of the wave, not the medium through which it passes), it follows that the wavelength will also be reduced to λ'. The reduction factor μ is called the *refractive index* of the medium. That is,

$$\mu = c/c' = \lambda/\lambda' \qquad (1.2)$$

If the amplitude of the wave is A, the most simple form of the wave equation in terms of a distance of propagation x is

$$a = A\cos(2\pi x/\lambda) = A\cos(2\pi x\bar{v}) \qquad (1.3)$$

1

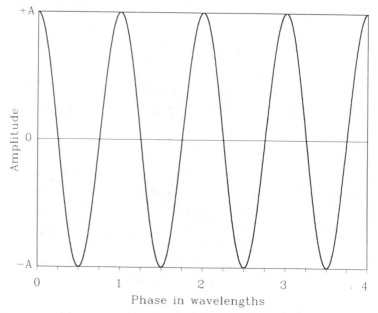

Figure 1.1. Portion of an electromagnetic wavetrain.

where we have defined a new frequency-related variable $\bar{v} = v/c$. The reason for doing this will become evident later. The wavetrain is illustrated in Fig. 1.1.

1.2. THE PRINCIPLE OF SUPERPOSITION

The *principle of superposition* states that light wave amplitudes are additive. If the wavetrains have the same frequency and a constant relative phase relationship (Fig. 1.2), the result is another cosine wave with a different phase and amplitude.

Let the wavetrains be

$$a_1 = A \cos(2\pi x_1 \bar{v}) \qquad \text{and} \qquad a_2 = A \cos(2\pi x_2 \bar{v})$$

Upon addition, we get

$$a = a_1 + a_2 = A[\cos(2\pi x_1 \bar{v}) + \cos(2\pi x_2 \bar{v})]$$
$$= 2A \cos(\pi \bar{v}\{x_1 - x_2\}) \cos(\pi \bar{v}\{x_1 + x_2\}) \tag{1.4}$$

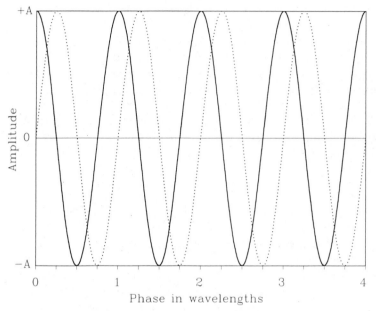

Figure 1.2. Superposition of two wavetrains.

Setting $x_1 = x_2 + \Delta$, and dropping the subscripts

$$a = 2A \cos(\pi\bar{v}\Delta) \cos(2\pi\bar{v}\{x + \Delta/2\}) \qquad (1.5)$$

which represents a wavetrain of the same frequency, but phase shifted by an amount $\Delta/2$ (second cosine term) and an amplitude (first cosine term) that depends on the phase difference. Observe that the amplitude can vary between $2A$ and zero.

1.3.   UNITS

The field of infrared spectrometry is sufficiently old for it to have accumulated a bevy of units, added to which are those specific to remote sensing, meteorology, and the like. Thus, before proceeding further, it is necessary to introduce the reader to some of the more important ones.

1.3.1. Spectrometric Units

The units most commonly used in spectrometry are the *nanometer* ($1\ nm = 10^{-9}\ m$) for wavelength and the *kayser* ($1\ k = 2.9979 \times 10^{10}\ Hz \approx 30\ GHz$) for frequency, although the word "kayser" is not widely used; usually substituted is the term *reciprocal centimeter* (cm^{-1}), often called (incorrectly) the *wavenumber*—not a unit, but a description.

 Other wavelength units that will be encountered are the *micrometer* (formerly called the *micron*; $1\ \mu m = 10^{-6}\ m$) and the *Ångstrom* ($1\ Å = 10^{-10}\ m$). The conversion between the two systems is

$$(\text{frequency in } cm^{-1}) = 10^{7}/(\text{wavelength in nanometers})$$

$$= 10^{4}/(\text{wavelength in micrometers})$$

1.3.2. The Units of Atmospheric Chemistry and Physics

Atmospheric chemists are interested in the fractional abundance of species, which they express as *parts per thousand/per million/per billion/per trillion* (‰, ppm, ppb, ppt). The fraction can be by mass (in which case the letter "M" should be appended (e.g., ppmM) or, more commonly, by volume (e.g., ppmV). The latter is more useful because it is numerically equal to the fractional number of molecules present in a given volume.

 Atmospheric remote sensing systems respond to the total number of molecules in the line of sight, otherwise known as the *column density*, most properly expressed as molecules·cm^{-2} (column). However, laboratory spectrometrists typically fill an enclosed cell of known length with gas at a particular pressure. Thus they often express their column densities as *centimeter·amagats*, where 1 amagat is the density at STP [standard temperature (273.15 K) and pressure ($1013.25\ mbar = 1.01325 \times 10^{5}\ N \cdot m^{-2} = 760\ mmHg$)]. The conversion is $1\ cm \cdot amagat = 2.68675 \times 10^{19}$ molecules·cm^{-2}.

 Regrettably, there's more:

- Results of field experiments are often quoted as *parts per million·meters* (ppm·m), that is, the product of a concentration and a path length. This "unit" can only be approximately converted because the actual number of molecules present depends on the local atmospheric density. Roughly, $1\ ppm \cdot m \approx 2.37 \times 10^{15}$ molecules·cm^{-2} at sea level.

- Meterologists specify the water vapor content of the atmosphere by *relative humidity*, the fractional degree of saturation; by *dewpoint*, the temperature at which the water vapor reaches saturation and condenses onto a surface; and by *precipitable millimeters*, the depth to which the water vapor would pool if it all condensed. Relative humidity and dew-point can be converted to number density (in molecules·cm^{-3}) through the use of tables (see the Bibliography), although for rough calculations relative humidity can be converted by the expression

$$\log_{10}[N_{H_2O}] = (RH)[17.1834 + 3.25067 \times 10^{-2}\Theta - 1.50311 \times 10^{-4}\Theta^2]$$

where N_{H_2O} is in molecules·cm^{-3}; RH = relative humidity; and Θ is the *celsius* temperature. Precipitable millimeters are directly convertible: 1 pr mm = 3.3423×10^{21} molecules·cm^{-2}.

1.4. TWO-BEAM INTERFERENCE

Amplitudes are required to understand how the superposition of wavetrains occurs. However, it is the square of the amplitude factor—the *intensity*—that provides the measurable physical phenomenon (at least at wavelengths shorter than the microwave region). For a justification of this step, consult any text on optics or electromagnetic theory.

Recall the expression (Eq. 1.5):

$$a = 2A \cos(\pi\bar{v}\Delta) \cos(2\pi\bar{v}\{x + \Delta/2\})$$

Then we obtain the intensity

$$I = a^2 = 4A^2 \cos^2(\pi\bar{v}\Delta) \tag{1.6}$$

Replacing A^2 by I_{in}, the incident intensity of each wavetrain, we get

$$I = 4I_{in} \cos^2(\pi\bar{v}\Delta) \tag{1.7a}$$

or

$$I = 2I_{in}[1 + \cos(2\pi\bar{v}\Delta)] \tag{1.7b}$$

Since $\bar{v}$ is a constant (and an inherent property of the wavetrains), it is evident that I will alternate between $4I_{in}$ and 0 as Δ changes through an amount $1/\bar{v}$ ($= \lambda$). These two extreme conditions are termed *constructive interference* and *destructive interference*, respectively, although any intermediate condition is equally possible.

In order for two-beam interference to occur, it is necessary to devise a means of generating two wavetrains having a fixed phase relationship. There are two basic approaches, both of which divide a single incoming wavetrain (of arbitrary phase properties) into two and introduce a phase shift after division. They are *division by wavefront* and *division by amplitude*. The former was historically the first approach (Thomas Young's double slit experiment of 1802, in which the phase difference was introduced by observing at different angles to the pair of slits). The second method is the topic of this book.

1.5. THE MICHELSON INTERFEROMETER

Amplitude division is usually performed through the use of a partially reflecting/ transmitting surface called a *beamsplitter*. This generates two in-phase wavetrains traveling in different directions. A phase shift is introduced into one or both beams, and the two are recombined at another (or the same) partially reflecting surface. One of the simplest optical systems that accomplishes this is the *Michelson interferometer* (Fig. 1.3).

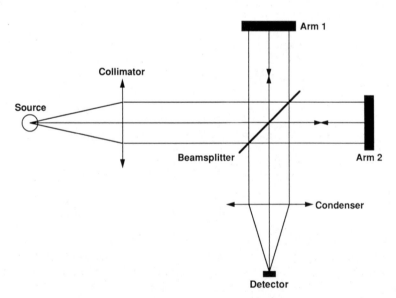

Figure 1.3. Principle of the Michelson interferometer.

1.5.1. Basic Implementation as a Spectrometer

Light from a point source (assumed monochromatic for the moment) is rendered parallel by a collimator (lens or mirror). The wavefront, within the limitations of geometrical optics, is now *plane*. The wavefront passes to a beamsplitter, here shown as an infinitely thin membrane (which can, in practice, be approximated) and is divided 50:50. The two resultant wavefronts travel to the plane mirrors and are reflected back on themselves. Provided that the mirrors are set perfectly perpendicular to the beams, the wavefronts on recombination at the beamsplitter (now acting as a recombiner) are still plane and parallel and will interfere by the principle of superposition. After focusing by a condenser, a detector will record an intensity that depends on the *path difference* that was imposed by the travel to and from the plane mirrors. That is, the distance of the mirrors from the beamsplitter is arbitrary—the only thing that matters is the *difference* in the paths.

Energy must be conserved, so what happens to the light not reaching the detector (assuming, for example, that the path difference was such as to cause destructive interference)? It is easily shown (and, indeed, intuitively obvious) that it returns to the source. Thus if the path difference changes such that the phase changes through π radians, the detector output will alternate between a maximum equal to the incoming intensity and zero.

Figure 1.3 presumed a static system. Suppose, however, that one arm moves at a constant velocity of V cm·s^{-1} (Fig. 1.4). When illuminated by a monochromatic point source, the detector will see a periodically varying signal and, from our previous analyses, we know that the output will be a cosine wave. The electrical frequency f of this waveform is readily calculated: the rate of change of path difference $d\Delta/dt$ is simply $2V$ cm·s^{-1}; so we have

$$f = 2\bar{v}V \text{ Hz} \tag{1.8}$$

remembering that the units of $\bar{v}$ are cm^{-1}. That is, a Michelson interferometer can be considered as a form of *frequency transducer* that converts optical frequencies (which are very high and well beyond the capability of known detectors to sense) down to electrical frequencies that can, in principle, have any value we choose since the mirror velocity V is controllable by the user. Furthermore, the transformation is *linear*: the amplitude of the electrical output is directly proportional to the incoming intensity.

It is these two properties that permit the Michelson interferometer to function as a spectrometer. Suppose that we replace the monochromatic source by a broad-band source (a tungsten lamp, for example). Such a source

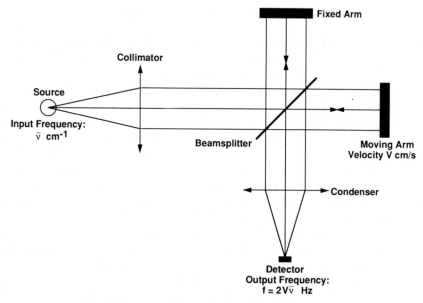

Figure 1.4. The Michelson interferometer as a spectrometer.

emits a wide range of optical frequencies that are independently (and uniquely) transformed into a range of electrical frequencies. The task now reduces to devising means of measuring the amplitudes of each of these electrical frequencies. If V is known, the input spectrum is immediately recoverable. In fact, a more elegant procedure is employed—the *Fourier transform* (whence the name of the technique).

1.5.2. Finite Field of View

Point sources do not exist. Real sources have a finite size and, as a result, light containing a finite range of angles traverses the interferometer. For example, if the source is circular with radius a and the collimator focal length is f, the range of angles traversing the system is simply a/f radians (rad). The impact on the interferometer can be described by simple geometry (Fig. 1.5).

Imagine that the condenser and detector in Fig. 1.3 are replaced by the eye. The beamsplitter acts as a partial mirror, so one will see the two plane mirrors superimposed and separated by some distance d. A ray entering at an angle θ is reflected at each plane mirror, as illustrated in the figure. The *physical*

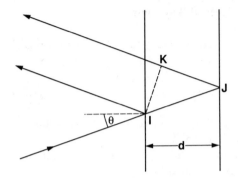

Figure 1.5. Geometric equivalent of a Michelson interferometer.

path difference is obviously the distance $IJ + JK$. From elementary trigonometry,

$$IJ + JK = d/\cos(\theta) + \cos(2\theta)\cdot d/\cos(\theta) = 2d\cos(\theta) \qquad (1.9)$$

Now, as we have seen, the path (phase) difference can also be described by the number of wavelengths n. That is, we can equate

$$n\lambda = 2d\cos(\theta) \qquad (1.10)$$

If the interferometer is embedded in a medium of refractive index μ, λ is reduced, so we must write

$$n\lambda = 2\mu d\cos(\theta) \qquad (1.11)$$

Finally, if we wish to eliminate wavelengths and work in frequency:

$$n = 2\mu\bar{v}d\cos(\theta) \qquad (1.12)$$

While n is a continuous variable, when it takes integer values the intensity will be *either* a maximum *or* a minimum, depending on the phase changes that occur on reflection at the beamsplitter. Which occurs is a matter for detailed electromagnetic theory but is of no concern here; as will become clear, the technique works equally well in either mode.

It is evident from Eq. 1.12 that the properties of a Michelson interferometer have a $\cos(\theta)$ angular dependence. This we shall see provides a connection between field of view and spectral resolution.

1.5.3. Finite Spectral Bandwidth

Suppose that the incoming monochromatic intensity to the spectrometer is F ($\equiv 2I_{in}$ in Eq. 1.7). Then Eq. 1.7b becomes

$$I(x) = F[1 + \cos(2\pi\bar{v}x)] \tag{1.13}$$

where we, from now on, use x (instead of Δ) to denote *path difference* and we explicitly recognize I as being a function of x.

Replacing the monochromatic source by the broad band source requires replacing F by $F(\bar{v})$ and integrating over frequency:

$$I(x) = \int_0^\infty F(\bar{v})[1 + \cos(2\pi\bar{v}x)]\, d\bar{v} \tag{1.14a}$$

$$= \int_0^\infty F(\bar{v})\, d\bar{v} + \int_0^\infty F(\bar{v})\cos(2\pi\bar{v}x)\, d\bar{v} \tag{1.14b}$$

The first integral is simply the total energy incident on the system and is therefore a constant. The second is related to the *cosine Fourier transform* of the incoming spectrum.

1.6. THE INTEGRAL FOURIER TRANSFORM

The definition of an *integral Fourier transform pair* is (in our notation)

$$G(x) = \int_{-\infty}^\infty F(\bar{v})\exp(2\pi i\bar{v}x)\, d\bar{v} \tag{1.15}$$

and

$$F(\bar{v}) = \int_{-\infty}^\infty G(x)\exp(-2\pi i\bar{v}x)\, dx \tag{1.16}$$

The proof of this assertion can be found in almost any textbook on analysis. Furthermore, because the spectrum $F(\bar{v})$ is (in the mathematical sense) *real*, the complex exponential in Eq. 1.15 can be replaced by a cosine:

$$G(x) = \int_{-\infty}^\infty F(\bar{v})\cos(2\pi\bar{v}x)\, d\bar{v} \tag{1.17}$$

The similarity of Eq. 1.17 to the second integral of Eq. 1.14b is obvious—only the integration limits differ. In fact, we are free to replace the lower limit of 0 in Eq. 1.14b by $-\infty$ because negative frequencies have no physical reality. In mathematical terms, $F(\bar{v}) \equiv 0$ for $0 \geqslant \bar{v} \geqslant -\infty$. Thus we may legitimately interpret the second integral of Eq. 1.14b as the cosine Fourier transform of $F(\bar{v})$.

What of the first integral in Eq. 1.14b? As we have already seen, it is a constant and, as we shall see further on, it is easily evaluated and subtracted from $I(x)$ *without any knowledge* of the nature of $F(\bar{v})$. That is,

$$G(x) = I(x) - C \tag{1.18}$$

where

$$C = \int_{-\infty}^{\infty} F(\bar{v})\,d\bar{v}$$

Under proper circumstances (to be discussed later), the inverse is also true:

$$F(\bar{v}) = \int_{-\infty}^{\infty} G(x)\cos(2\pi\bar{v}x)\,dx \tag{1.19}$$

Since $G(x)$ is directly related to the measured output, called an *interferogram*, of the spectrometer (now to be called a *Fourier transform spectrometer*, or FTS for short), it follows that if we can solve Eq. 1.19 we have devised a means of recovering $F(\bar{v})$, the original input spectrum and the object of the exercise.

Unfortunately, we cannot solve Eq. 1.19; at least not quite. First and foremost, the use of an integral sign implies that we have knowledge of x at infinitesimally small intervals (i.e., an infinite number of samples). As we shall soon see, we can have (and only need) knowledge of x at discrete intervals, which will entail replacing the integral by a summation but thereby rendering the problem tractable on a computer. Second, and equally important, the interferogram is produced by a real instrument and can be generated only over some finite range of path difference x (0 to L, say). What are the implications of this?

Knowing $G(x)$ only over some finite range is exactly equivalent to multiplying a theoretically infinite version of $G(x)$ by a function that takes the value 1 between $x = 0$ and $x = L$ and is 0 everywhere else. Such a function is usually called a *boxcar* function. Thus what we actually will try to solve is not Eq. 1.19

but a modified version:

$$F'(\bar{v}) = \int_{-\infty}^{\infty} G(x) \cdot B(x) \cos(2\pi\bar{v}x)\, dx \qquad (1.20)$$

where $B(x) = 1$ for $0 \leqslant x \leqslant L$; 0 otherwise. What we must now try to resolve is the relationship between $F(\bar{v})$, the answer we really want, and $F'(\bar{v})$, the answer we are actually going to get. The key lies in the procedure known as *convolution*.

1.7. CONVOLUTION

The formal definition of convolution is

$$c(t) = \int_{-\infty}^{\infty} g(\tau) \cdot h(t - \tau)\, d\tau \qquad (1.21)$$

often written in a shorthand form as $c = g \otimes h$. Now, as is shown in many

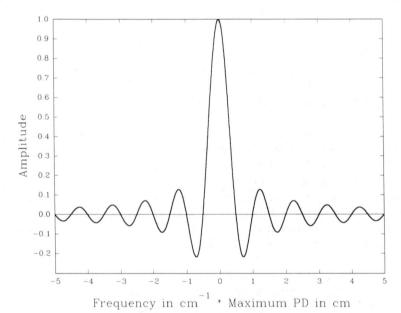

Frequency in cm^{-1} * Maximum PD in cm

Figure 1.6. The sinc function.

textbooks (see the Bibliography), if we represent the Fourier transforms of c, g, and h by $C(f)$, $G(f)$, and $H(f)$, then

$$C(f) = G(f) \cdot H(f) \tag{1.22}$$

from which is immediately follows that $c = h \otimes g$ also. The required relationship between $F(\bar{v})$ and $F'(\bar{v})$ is therefore

$$F'(\bar{v}) = F(\bar{v}) \otimes J(\bar{v}) \tag{1.23}$$

where $J(\bar{v})$ is the Fourier transform of the boxcar function $B(x)$. It is easily demonstrated that, with appropriate normalization,

$$J(\bar{v}) = \sin(2\pi\bar{v}L)/(2\pi\bar{v}L)$$
$$= \text{sinc}(2\pi\bar{v}L) \quad \text{in shorthand notation} \tag{1.24}$$

where L is the maximum path difference. The function is plotted in Fig. 1.6.

1.7.1. Physical Interpretation

The impact of the sinc function on the "ideal" spectrum $F(\bar{v})$ is to "smear out" the fine structure. That is, any spectral structure significantly sharper than the width of the sinc function will be lost. This is the definition of *spectral resolution*—the measure of the ability of a spectrometer to reproduce the incoming spectrum.

Another way of looking at convolution is to imagine the sinc function as an *averaging* function. A similar role is played by the slit in a scanning spectrometer or by the finite size of the detector elements in an imaging spectrometer. That is, any real spectrometer of any type has a finite resolution. It is conventional to define the resolution as the full width at half height of the main peak. For an FTS, this occurs at a frequency shift from the peak of $0.30168/L$ cm^{-1} (L = maximum path difference in centimeters. Thus the spectral resolution of an FTS is twice this value ($0.6034/L$ cm^{-1}). Most importantly, the resolution is seen to be *inversely proportional to the maximum path difference*.

It appears therefore that there is no fundamental limit to the spectral resolution of an FTS and, in a sense, this is true. Laboratory systems exist in which L can be as great as 600 cm. In fact, the real limitation turns out to be (anticipating a later result) the fact that the signal-to-noise ratio declines at least as fast as resolution, so most real FTS are bounded not by mechanics and optics but by other considerations.

A point not yet touched upon is the obvious fact (see Fig. 1.6) that the sinc function shows negative peaks whose amplitude dies away quite slowly. This turns out to be a problem only if certain not strictly legitimate operations are performed on the output spectrum. It is, however, possible to play numerical games with the interferogram that "damp out" the oscillations. The procedure is called *apodization* (literally "feet removal"!), but there is a price: any such procedure widens the central peak. Apodization always results in a loss of spectral resolution and can, in addition, introduce some undesirable "cross talk" into the spectrum.

1.8. THE DISCRETE FOURIER TRANSFORM

The *discrete Fourier transform* is actually a different operation from the integral transform and rather more difficult to understand. In one sense, integral transforms deal with arbitrary functions. Discrete transforms are more applicable to *periodic functions* or, as we shall see, functions that can be made to appear as if they *were* periodic (in short, we cheat!). The "cheating" is necessary because, as we saw earlier, the integral transform cannot be solved with real-world data.

Suppose that the interferogram is sampled at exactly equal intervals δx (how δx is chosen is a major topic of Chapter 4), with the positions set such that the first sample is acquired at $x = 0$ (the condition of *zero path difference*, ZPD). If the maximum path difference is L, then we shall acquire $N = 1 + L/\delta x$ samples. It transpires (reasonably enough) that the output of the discrete transformation is a set of N spectral intensities equally spaced in frequency— say, $\delta \bar{v}$. The formalisms are

$$I(m\,\delta x) = \sum_{k=0}^{N} F(k\,\delta\bar{v})\exp(2\pi i k\,\delta\bar{v}\cdot m\,\delta x) \tag{1.25}$$

and

$$F(n\,\delta\bar{v}) = \frac{1}{N}\sum_{j=0}^{N} I(j\,\delta x)\exp(-2\pi i n\,\delta\bar{v}\cdot j\,\delta x) \tag{1.26}$$

although they are rarely used in precisely this form (but note the renormalization of Eq. 1.26). By the use of clever factorization, it is possible to reduce the computational burden (so-called fast Fourier transforms) to the point that with quite modest computers we can now transform more than a million samples (actually $2^{20} - 1$) in a few seconds. For expositions of this topic, see the Bibliography.

CHAPTER

2

THE IDEAL FOURIER TRANSFORM
SPECTROMETER

We have already met the simplest form of FTS—the Michelson interferometer (Fig. 1.4).

While this configuration is very useful for explaining the properties of an FTS, it is not necessarily the ideal optical arrangement—although, for example, the *Voyager* IRIS FTS looks remarkably like Fig. 1.4. Some alternative approaches will be covered in Chapter 4.

In Chapter 2, we concentrate on the properties of an idealized system and compare them to an equally idealized dispersive spectrometer since we must also address the issue of why one would choose to use such a seemingly complex approach to spectrometry as the FTS.

2.1. ÉTENDUE

The French term *étendue* (which translates as "extent") was appropriated many years ago to describe the most fundamental constant of an optical system—the product of the area A of the light beam and the solid angle Ω contained within the beam (the $A\Omega$ product). While the term "throughput" will sometimes be encountered for this parameter, it is better employed for the product of étendue and overall system transmittance (namely, $A\Omega\tau$). The importance of étendue lies in the fact that it is one of the major controlling factors in the determination of the signal-to-noise ratio. Furthermore, it is a design parameter: it can be maximized by a proper choice of configuration. The other major control—the system transmittance—is much more at the mercy of technology (available mirror coatings, filters, material transparency, etc.).

Étendue is a property that is *at best* conserved through an optical train. That is, the *system* étendue will be that of lowest value in any part of the system. No amount of optical "trickery" will change this fact. The geometry of étendue is shown in Fig. 2.1.

15

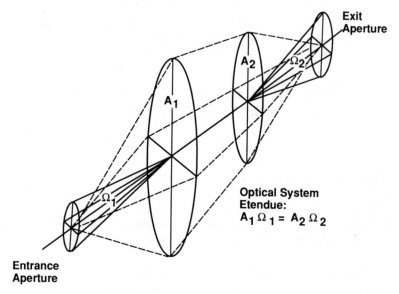

Figure 2.1. The definition of étendue.

It is usual to make the most expensive (= most critical) optical element the controlling factor. For a dispersive spectrometer, this will often be the prism or grating. For an FTS, it is likely to be the beamsplitter.

2.1.1. Beam Area

The area of the beam passing through the FTS is primarily a function of the largest available (or permissible, in size-constrained systems) piece of transparent material for the beamsplitter. While our "idealized" system shows the beamsplitter as an infinitely thin membrane (and, indeed, systems using unsupported thin film beamsplitters have been built), it is much more common for the beamsplitter to be made from a single piece of crystalline or glassy material. As we shall see, the major area of use of FTSs is in the infrared where conventional glasses are opaque. Beamsplitters therefore must be made from relatively exotic materials such as single crystals of calcium fluoride or potassium bromide. Such crystals are *very* expensive. As a matter of practicality, a 5- to 10-cm diameter would be a typical size.

In the worst case of a beamsplitter of area S used at $45°$ (note that, as we shall see, there is no fundamental requirement that the arms of an FTS be mutually perpendicular), the beam area $A = S/(2^{1/2})$.

2.1.2.   Solid Angle

As is true of all types of spectrometer, the permissible solid angle contained by the light passing through an FTS is intimately connected to the desired spectral resolution $\delta\bar{\nu}$. We begin with Eq. 1.11:

$$n\lambda = 2\mu d \cos(\theta)$$

On-axis, some wavelength λ_0 is transmitted. At a small angle θ off-axis, a different wavelength λ_1 is seen. Thus

$$n\lambda_0 = 2\mu d \qquad \text{and} \qquad n\lambda_1 = 2\mu d \cos(\theta) \approx 2\mu d(1 - \theta^2/2) \tag{2.1}$$

Substituting, we get

$$n\lambda_1 = n\lambda_0(1 - \theta^2/2)$$

Setting $\lambda_0 - \lambda_1 = \delta\lambda$ and dropping the subscripts, we obtain

$$\delta\lambda/\lambda = \theta^2/2 \tag{2.2a}$$

Differentiation of Eq. 1.1 reveals that $\delta\lambda/\lambda = \delta\bar{\nu}/\bar{\nu}$. Thus it is also true that

$$\delta\bar{\nu}/\bar{\nu} = \theta^2/2 \tag{2.2b}$$

Now θ is the off-axis angle, so it defines the semiangle of the solid angle cone subtending Ω steradians (sr). By standard solid geometry,

$$\Omega = \pi\theta^2 \tag{2.3}$$

and we finally arrive at

$$\delta\lambda/\lambda = \delta\bar{\nu}/\bar{\nu} = \Omega/(2\pi) \tag{2.4}$$

This is a most important result and one to which we shall make reference in the future. Finally, then

$$(A\Omega)_{\text{FTS}} = 2^{1/2}\pi S/(\delta\lambda/\lambda) \tag{2.5}$$

As a reminder, the *system étendue cannot exceed* the value of $A\Omega$ defined by the projected area of the beamsplitter but can (with bad design!) be less.

2.1.3. Comparison to Dispersive Systems

A diffraction grating spectrometer can be shown to provide a greater étendue than one using a prism (Jacquinot, 1954), so prism spectrometers will not be further discussed. It is also true that, for best efficiency, a diffraction grating should be used as closed to its blaze angle ϕ as possible. At this condition

$$n\lambda = 2b\sin(\phi) \tag{2.6}$$

where n is the diffraction order and b is the width between adjacent rulings.
    The angular dispersion is

$$\delta\phi = [n\delta\lambda]/[2b\cos(\phi)] = \tan(\phi)\cdot(\delta\lambda/\lambda) \tag{2.7}$$

In the plane perpendicular to the dispersion, the allowable angle is less easily calculated because it depends on the type of source, the type of detector, and, above all, on optical aberrations. However, systems have been reported for which the angular slit length β was as much as 1/30, so we shall assume this to be the case here.
    Now

$$\Omega = \delta\phi\cdot\beta = \beta\tan(\phi)\cdot(\delta\lambda/\lambda) \tag{2.8a}$$

and, if the area of the grating is S,

$$A = S\cos(\phi) \tag{2.8b}$$

whence

$$(A\Omega)_{\text{grating}} = S\beta\sin(\phi)\cdot(\delta\lambda/\lambda) \tag{2.9}$$

For a given surface area and spectral resolution, then, we find

$$\frac{(A\Omega)_{\text{FTS}}}{(A\Omega)_{\text{grating}}} = \frac{2^{1/2}\pi}{\beta\sin(\phi)}$$

The absolute (and unrealistic) maximum value for $\sin(\phi)$ is 1 and we have already seen that β is unlikely to be greater than 1/30, so the advantage of an FTS over a grating (in terms of étendue) can be as large as a factor of 150 or more for the same size. Now it is certainly true that gratings can be made much larger than beamsplitters. For a price, a useful area of $60 \times 60\,\text{cm}$ can be achieved (with a concomitantly interesting optical system to feed it), whereas an FTS would generally be limited to about 10 cm. Thus, in principle, an immense grating spectrometer need be no more than 5 times worse (in terms of étendue) than a physically much smaller FTS of the same resolution. This statement is not intended to imply that dispersive spectrometers are worthless, because étendue is by no means the entire story. There are certain conditions wherein a dispersive system can, in fact, outperform an FTS. We shall meet some of them later.

The original realization of this benefit (which is shared by another type of interferometer called a Fabry–Perot interferometer) belongs to Pierre Jacquinot, so the gain is often called the *Jacquinot advantage* of an FTS.

2.2. MULTIPLEXING

The term *multiplexing* was coined by communications engineers to describe techniques for transmitting multiple, simultaneous signals through a single communications channel (multiple telephone conversations through a single wire, for example). There are two types: *time-division multiplexing*, in which the signals are successively sampled, sent through the channel, and reconstructed at the end (telephone systems use this method); and *frequency-division multiplexing*, in which the signals are impressed on carriers of different frequencies. Radio or TV broadcasting is an example of this method; an FTS is another.

2.2.1. Implementation

At the time that Fourier transform spectrometry was first being developed (1955–1965), the only alternative approaches for infrared spectrometry were dispersive systems using a single detector that observed the spectrum point-by-point in a sequential fashion (usually by rotating the prism or grating). It was recognized that this was exceedingly inefficient as compared to, for example, the photographic spectrographs used in visible and ultraviolet spectrometry in which *all* the spectrum was observed *all* the time (questions of quantum efficiency aside). It early became evident that Fourier transform spectrometry shared this benefit of simultaneous observation and, in particular, since each optical frequency is converted to a different electrical frequency (Eq. 1.8), that it did so by frequency-division multiplexing. This particular benefit of an FTS is often called the *Fellgett advantage* after Peter Fellgett, the investigator who first recognized it. The resultant wide simultaneous frequency (wavelength) coverage available with an FTS is one of the major reasons for its current dominance in most forms of infrared spectrometry.

The multiplexing gain of an FTS over a sequentially scanning system can easily exceed 1000, and factors as large as 10^6 have been reported in the literature (Connes et al., 1970).

2.2.2. A Warning

The multiplexing advantage applies only if the signals themselves can be considered to be noise free (that is, the only significant noise is contributed by the detector or electronics). If this is not true (as in the visible and ultraviolet) the benefit disappears. This topic will be treated in some detail in Chapter 4.

2.2.3. Terminology

As more and more dispersive systems are being built using arrays of detectors, some loose terminology has resulted. It is often stated that array detector systems have a "multiplex advantage." They do not. They have a *multichannel advantage.* Why does it matter when the outcome—simultaneous observation of many wavelengths—is the same? It matters because FTSs can also be used with detector arrays (discussed later in this chapter). Thus an FTS can display *both* advantages, and it is important to distinguish between them.

2.3. SPECTRAL RESOLUTION AND INSTRUMENTAL LINE SHAPE

In Chapter 1 we met the most common definition of spectral resolution: the width of the instrumental line shape (ILS, sometimes called the *impulse response*) at half height. Other definitions do exist (the Rayleigh criterion, for example), but they will not concern us here. The important thing is to have a definition and to use it consistently.

2.3.1. Spectral Resolution

We have already seen that the spectral resolution of an FTS is essentially unlimited:

$$\delta\bar{v} \approx 0.6/L \qquad (2.10)$$

where L is the maximum path difference used.

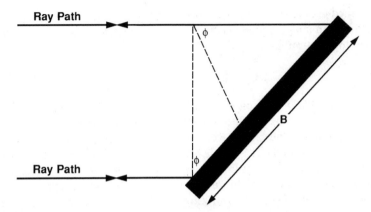

Figure 2.2. Geometry of a diffraction grating (Littrow condition).

For a diffraction grating (which is actually a species of multiple-beam interferometer—each facet or ruling is the source of one of the beams) the maximum possible resolution is "built in" and controlled by the path difference between the extreme facets. Figure 2.2 shows the geometry for the optimum near-Littrow condition. For a grating of size B (in the direction perpendicular to the ruling) and blaze angle ϕ, the geometric path difference between extreme rays is seen to be $B\sin(\phi)$.

At a wavelength λ we can equate

$$m\lambda = B\sin(\phi) \tag{2.11a}$$

A wavelength $(\lambda + \delta\lambda)$ will be resolved when

$$(m - \tfrac{1}{2})(\lambda + \delta\lambda) = B\sin(\phi) \tag{2.11b}$$

Substituting for m, rearranging, and dropping the negligible term $\delta\lambda/2$, we obtain

$$\delta\lambda/\lambda \approx \lambda/[2B\sin(\phi)] \tag{2.12}$$

But $\delta\lambda/\lambda = \delta\bar{v}/\bar{v}$ and $\bar{v}\lambda = 1$, so we finally get

$$\delta\bar{v} = 1/[2B\sin(\phi)] \tag{2.13}$$

In fact, a grating is rarely used at its limit. The resolution is commonly determined geometrically by the width of the entrance and exit apertures (the detectors in an imaging spectrometer). The previous calculation merely provides a lower (diffraction) limit.

2.3.2. Instrumental Line Shape

We have seen that the ILS of an FTS is analytic—a sinc function in the simplest case. The ILS of a dispersive spectrometer can be shown to be roughly triangular (or trapezoidal in some cases). "Roughly" is, unfortunately, the word. Although the measurements of the ILS of a dispersive spectrometer are not easy to make, it is found that the deviations from triangularity can be substantial—asymmetry being a very common condition. Furthermore, because the deviations are largely a function of skew rays striking the disperser, additionally aggravated by diffraction and optical aberrations, the ILS becomes markedly frequency (wavelength) dependent. The ILS of an FTS is constant (on a frequency scale). Again, this property has a name—the *Connes advantage* of an FTS.

2.3.3. Consequences

In Chapter 3, we shall meet some of the foundations of atmospheric spectra and spectrometry. For now, two specific properties of atmospheric spectra are pertinent. First, their amplitudes are calculable to high precision on an absolute scale (to better than 1% in favorable cases), so one of the preferred ways of analyzing atmospheric spectra is through iterative modeling procedures: when the model matches the observed data, the content of the data bases generated in the modeling process contain all the information extractable from the spectrum (debates about uniqueness aside). Second, it is a property of atmospheric absorptions and emissions that the spectral features due to individual transitions (for historical reasons called *lines*) show, at a given temperature and pressure, almost the same width on a frequency scale (within a factor of 2–3). The fact that the ILS of an FTS is both constant and analytic is what makes the modeling process feasible. To do the same thing with a dispersive spectrometer involves a massive undertaking of empirical characterization of the ILS to the point that it is rarely done for more than very limited spectral ranges.

2.4. FREQUENCY ACCURACY

A key factor in the analysis of all types of spectra is *frequency accuracy*. It is particularly important for atmospheric spectrometry because the infrared region of relevance to remote sensing (roughly $1-1000\,\mu m$; $10-10,000\,cm^{-1}$)

contains several hundred thousand identifiable lines. Without accurate frequencies, identification becomes difficult or impossible.

2.4.1. Fourier Transform Spectrometers

The Michelson interferometer was originally devised as a means of determining absolute wavelengths [by comparing the path difference to a standard meter bar and counting how many modulation periods (*fringes*, as they are often called) were seen]. While such use is no longer common, the property of accurate wavelength and frequency determination remains. The key is to measure the path difference *very* accurately (to a few atomic diameters!). The method universally employed is to feed the FTS with a monochromatic beam of accurately known frequency (often a frequency-stabilized red helium-neon laser) simultaneously with the external infrared beam. The control beam can be very small and is, in any case, normally at a much shorter wavelength than the region of interest, so separating the two is straightforward. Through the use of a relatively simple control system (outlined in Chapter 4), the "unknown" frequencies can be determined to very high accuracy (1 part in 10^8 is quite common).

Naturally, real life is not quite so simple. There are a number of subtle effects that must be taken into account to preserve the potential accuracy but they are generally well understood, and it is no coincidence that the best current laboratory measurements of line frequencies for atmospheric molecules are invariably taken using FTSs. One not-so-subtle effect that must be taken into account in spaceborne remote sensing is the Doppler effect generated by the high speeds at which spacecraft move ($\sim 7\,\mathrm{km\cdot s^{-1}}$ for a near-Earth orbiter; considerably more for a planetary flyby). If the velocity component in the line of sight is v, then the relative shift of spectral features with respect to laboratory standards is

$$\delta\bar{v}/\bar{v} = v/c \qquad (2.14)$$

where c is the velocity of light. The precision of an FTS is such, however, that the effects of wind shears of a few meters per second can be discerned in high-resolution stratospheric spectra, even when superimposed on much larger orbital Doppler shifts.

2.4.2. Dispersive Spectrometers

From the standpoint of frequency/wavelength determination, dispersive spectrometers make strictly *relative* measurements. That is, they must be provided

with a suite of wavelength calibration sources because the wavelength falling
on any given detector is primarily a function of optical alignment and stability.
For remote sensing systems, this can be a significant problem.

2.5. MODULATION EFFICIENCY

So far, we have concentrated on features of an FTS where it excels over dispersive
spectrometers. There is, however, a property of an FTS that has no counterpart
in dispersive systems. It is called *modulation efficiency*.

When we discussed two-beam interference in Chapter 1 (Eqs. 1.6 and 1.7),
we explicitly assumed that the amplitudes of the two interfering beams were
identical. What if they are not equal? Looking back at Fig. 1.3, we observe
that each beam is reflected once and transmitted once at the beamsplitter. Let
the intensity reflection and transmission coefficients be R and T, respectively.
Replacing I_{in} in the expressions, we find the multiplying factor to be M, where

$$M = 4RT \qquad (2.15)$$

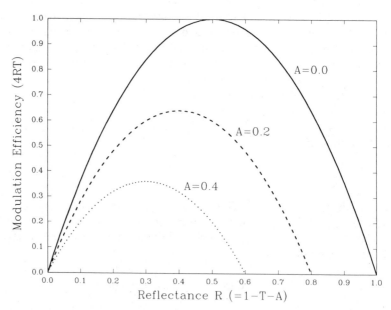

Figure 2.3. Modulation efficiency as a function of beamsplitter reflectance and absorptance.

In an ideal case, $R = T = 0.5$, whence $M = 1$. Unfortunately, real beamsplitters rarely meet this requirement and then only at one or two wavelengths. Fortunately, *provided that the beamsplitter and its coatings are lossless*, M is relatively insensitive to changes in R and T (Fig. 2.3). For example, for $R = 0.4$ and $T = 0.6$ (or vice versa), $M = 0.96$, a relatively small loss. An absorbing beamsplitter is more of a problem. Suppose we have $R = T = 0.3$ (implying a 40% loss in the beamsplitter, a not impossible case). Then we get $M = 0.36$, a major reduction in efficiency.

2.5.1. The Impact of Impaired Modulation

An FTS lives and dies by the modulation it impresses on the incoming signals. If M goes to zero, so does the signal. The impact can be even more serious, as we shall see in Chapter 4. If the photon shot noise of the source is significant, the signal is reduced *without reducing the noise* because although the signal modulation may be reduced, unmodulated photons are still passing through the system. It is therefore crucial to obtain the best possible beamsplitter and coatings. As always, this costs money and is a major reason why this element would usually be, by a large factor, the most expensive optical element in the FTS.

The coatings are not the only way an FTS can loose modulation efficiency. Without going into details, optical aberrations, mirror surface defects, and misalignments can also cause losses. Control of such effects is one of the many reasons why FTS design is still somewhat of an art form.

2.6. IMAGING MODES

The advent of detector arrays has enabled a new type of spectrometer—the *imaging spectrometer* that can generate images in much narrower bands (e.g., AVIRIS) than was possible using band-pass filters (e.g., LANDSAT Thematic Mapper). Both FTS and dispersive spectrometers can be used in this mode.

2.6.1. FTS Implementation

The "standard" configuration for an FTS is to use a single detector on-axis whose maximum angular subtense was given by Eq. 2.4:

$$\delta\bar{v}/\bar{v}_{max} = \Omega_{max}/(2\cdot\pi)$$

where $\bar{v}_{max}$ is the highest frequency (shortest wavelength) of interest. This criterion was dictated by the requirement that the change in frequency owing to the off-axis angle not degrade the spectral resolution. Typically, then, one would prefer that the detector subtense not exceed some 50–70% of Ω_{max}. With single-detector systems, it is usual to image the system exit pupil onto the detector because better illumination uniformity—and hence less noise—results.

In order to employ an imaging mode, however, this stratagem is not available unless a "fly's-eye" condenser system can be implemented. Typically, then, the remotely sensed scene is imaged onto the detector elements, whose individual dimensions therefore define a *ground sampling distance* (GSD). We shall now investigate the coupling between this parameter and the spectral resolution.

An FTS is usually considered to be "nondispersive." This is nearly, but not quite, true. Remembering Eq. 1.12,

$$n = 2\mu\bar{v}(\theta)x\cos(\theta)$$

we see that, going in any direction from the optic axis, a change in frequency (for a given order n) will occur. It is perhaps clearer if we rewrite Eq. 1.12 as

$$\bar{v}(\theta) = (2\mu x/n)\sec(\theta)$$

and, recognizing that the on-axis frequency $\bar{v}(0) = 2\mu x/n$, we get

$$\bar{v}(\theta) = \bar{v}(0)\sec(\theta) \tag{2.16}$$

Differentiating, we get the angular dispersion of an FTS:

$$\delta\bar{v}/\delta\theta = \bar{v}(0)\sec(\theta)\cdot\tan(\theta) \approx \bar{v}(0)\cdot\theta \qquad \text{for small } \theta \tag{2.17}$$

which, while small, is not zero. A critical point to note, however, is the radial symmetry: the dispersion is the same in any direction from the axis. This is in complete contrast to a grating or prism, whose dispersion is in only one direction.

The radial symmetry of an FTS offers us the opportunity of installing a two-dimensional detector array in the focal plane and using it directly as a spectral camera. The requirement is that, on any detector element, the angular dispersion be less than the desired resolution. This is essentially the same requirement as the on-axis case, but generalized for off-axis detector elements (often called *pixels*).

Suppose that the required GSD is G meters and the range from the spectrometer to the surface (or "target," in general, which could be the Earth's limb) is R meters. The angular subtense of the GSD at the sensor entrance pupil is therefore $\delta\phi = G/R$.

The change in angles through an optical system is termed its *magnification* m. Thus the internal subtense $\delta\theta$ is

$$\delta\theta = m\,\delta\phi = mG/R \qquad (2.18)$$

Similarly, if the coverage requirement is C meters and recognizing that this is centered on the optic axis, the maximum off-axis angle Θ is

$$\Theta = mC/(2R) \qquad (2.19)$$

The requirement can therefore be written as

$$\delta\bar{v} \geq \bar{v}_{max}m^2CG/(2R^2) \qquad (2.20)$$

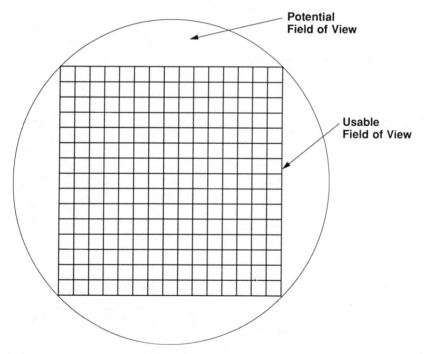

Figure 2.4 Matching a square detector array to an FTS.

The largest permissible $N \times N$ detector array, then, becomes a function of the required spectral resolution, the highest frequency of interest, and the observation geometry:

$$N = C/(2^{1/2} G) \leqslant 2^{1/2} (\delta \bar{v}/\bar{v}_{max})(R/[mG])^2 \qquad (2.21)$$

where the extra $2^{1/2}$ comes from the fact that the off-axis angle must be measured along the diagonal of a square array (Fig. 2.4). This is an interesting result, because the only parameter entering the equation that pertains to the optical system is m, the magnification. This can be contrasted to dispersive imaging spectrometers, which have quite different criteria.

2.6.2. Another Warning

Since an FTS generates its information through the amplitude of the individual elementary cosine waves that make up the interferogram, it follows that any additional amplitude modulation would be fatal. Such modulation can occur if there is any image motion, pointing jitter, or significant atmospheric tubulence and the target under investigation is nonuniform. In general, an imaging FTS must be used in a so-called staring mode: the system must point accurately and steadily at the target for the time it takes to acquire the interferogram.

2.6.3. Dispersive Spectrometer Implementation

A dispersive imaging spectrometer can only support one dimension of imaging (along the slit) because the other dimension is used for the spectral dispersion

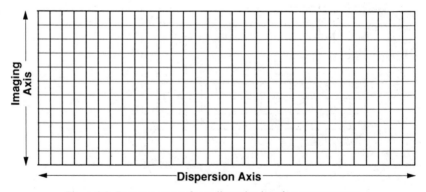

Figure 2.5. Detector arrays for a dispersive imaging spectrometer.

(Fig. 2.5). Images are built up by making successive exposures so that the image lines are "stacked" side by side. This can either be done using the motion of the aircraft or spacecraft (*push-broom imaging*) or by the use of a sideways-scanning mirror (*whisk-broom imaging*). Both methods are widely used, though in either case there is a strong coupling between exposure time and the speed/altitude of the platform. Note that the conditions are quite the opposite to an FTS: image motion is *required* for a dispersive system.

One consequence of the decoupling of imaging and spectrometry is that rectangular detector arrays can be used. The limitations on the number of elements in each dimension are optical aberrations, unlike an FTS.

2.6.4. Discussion

Which, then is the better: a dispersive spectrometer or an FTS? As in most things, the answer depends on what one is trying to accomplish. There is no doubt that for a large proportion of infrared investigations, the FTS is a clear-cut winner (sometimes by factors of hundreds to thousands). As we shall learn in Chapter 4, however, if the photon statistics of the source or the background are significant, the choice becomes less obvious. Indeed, one of the purposes of this book is to provide the reader with some of the background needed to make an informed choice.

THE PHYSICS AND CHEMISTRY
OF REMOTE SENSING

In this chapter we shall discuss the nature of the signals that occur in remote sensing. While the bias is certainly toward atmospheres, any downlooking remote sensor necessarily sees the surface if the atmosphere is sufficiently transparent; a proper description of the signals must include the contributions from both. This is the field of *radiative transfer*. The absorptions and emissions of the atmosphere and surface are the realm of *spectrometry* and *atmospheric chemistry*. Atmospheric spectra, in particular, are strongly influenced by the conditions of temperature and pressure under which they are formed. This is the purview of *atmospheric physics*. Indeed, one of the major current uses of atmospheric remote sensing is the routine determination of global atmospheric temperature and humidity profiles, which are themselves primary inputs to models for numerical weather forecasting and, in the long term, climatology.

3.1. THE PHYSICS OF ATMOSPHERES

The most variable property of an atmosphere is its vertical temperature profile, varying with altitude (Fig. 3.1), time, and place. At the surface, seasonal and diurnal temperature changes of 10% are not uncommon. By contrast, the surface pressure rarely changes by more than 2–3%.

3.1.1. The Atmospheric Temperature Profile

As can be seen in Fig. 3.1 (which is a worldwide grand average profile called the *1976 U.S. Standard Atmosphere*), the atmosphere conveniently divides itself into vertical regions. The atmospheres of other planets show similar (but far from identical) divisions.

Very close to the surface (too close to show on the figure) is the *mixing layer*, typically 1–2 km thick. In this region turbulent mixing occurs, so it plays a major role in the coupling of the atmosphere and the surface.

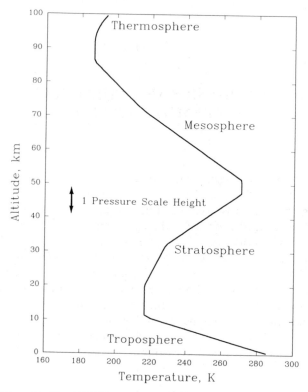

Figure 3.1. The 1976 U.S. Standard Atmosphere temperature profile.

The region from the surface to some 10–15 km (very variable with time and place) is the *troposphere*, a region in which the temperature declines with altitude at a rate (the *lapse rate*) very close to that predicted by the cooling of a parcel of atmosphere rising and cooling adiabatically (the phenomenon of *convection*). The rate is about 6.5 K per kilometer.

After passing through a region of near-constant temperature (the *tropopause*), we enter the *stratosphere*, a region of increasing temperature primarily caused by solar heating of ozone (which is strongly concentrated into the stratosphere).

At a higher altitude we pass through the *stratopause* into the *mesosphere*, where the atmosphere again cools but at a slower rate than the troposphere (which inhibits convection).

Beyond the *mesopause* is the *thermosphere*, where the temperature begins to rise again as all surviving molecules dissociate. At the very low densities encountered in the thermosphere, the very concept of a temperature becomes debatable because collisions between molecules become very infrequent (by comparison to the time required to emit or absorb a photon). We are in the realm were *local thermodynamic equilibrium* (LTE) fails.

3.1.2. The Atmospheric Pressure and Density Profile

The pressure at any level in the atmosphere is simply the weight of the total atmospheric column above it. Obviously, then, the pressure must decline with altitude. To a reasonable approximation, the atmosphere is said to be in *hydrostatic equilibrium*, although were this to be truly the case there would be no weather! Within the limits of the assumption,

$$dP/dz = -gNM \tag{3.1}$$

where P is the pressure, z is the altitude, g is the acceleration of gravity (~ 981 cm·s^{-2} at the surface), N is the number of molecules per unit volume, and M is the mean molecular mass (~ 29 for the Earth's lower atmosphere).

Similarly, the atmosphere behaves almost as a perfect gas:

$$P = NRT \tag{3.2}$$

where T is the absolute temperature and R is a gas constant ($= 1.363 \times 10^{-22}$ for P in standard atmospheres and N in molecules·cm^{-3}).

Combining Eqs. 3.1 and 2 yields

$$dP/P = -dz/[(RT)/(Mg)] \tag{3.3}$$

The term in square brackets has the dimensions of a length and is called the *scale height*, usually denoted by H. Although H is a function of temperature, composition, and altitude, for the Earth's lower atmosphere it is roughly 8 km. We shall encounter H again.

Solving Eq. 3.3 and setting the surface pressure equal to P_0, we get

$$P(z) = P_0 \exp\left(-\int_0^z dz/H\right) \tag{3.4a}$$

which, for a simple isothermal atmosphere (H constant), reduces to

$$P(z) = P_0 \exp(-z/H) \tag{3.4b}$$

We can therefore interpret H as the altitude range over which the pressure declines by a factor $1/e$. The horizontal dotted lines in Fig. 3.1 record this interpretation.

3.1.3. Model Atmospheres

A tabulation of atmospheric temperature, pressure, and density vs. altitude is called a *model atmosphere*. The U.S. Standard Atmosphere is such a tabulation; there are many others available that provide average conditions for a variety of locations. Model atmospheres are a crucial data base for remote sensing because they permit the computation of the expected transmission through the atmosphere. Furthermore, since the actual atmosphere along any given line of sight will differ from the models, the models are often used as a starting point for the *retrieval* (i.e., measurement) of the atmospheric properties.

3.2. THE CHEMISTRY OF ATMOSPHERES

The recent publicity given to anthropogenic modifications of the atmosphere, for example—

- The increase in "greenhouse" gases that may lead to global warming
- The impact of manufactured chlorofluorocarbons (CFCs) on stratospheric ozone
- Urban and regional atmospheric pollution episodes
- Sulfur and nitrogen oxides contributing to acid precipitation

have given a major impetus for the development of new measurement systems (both remote and *in situ* sensors) and climate models to quantify these phenomena. Of particular importance is the fact that gases injected into the atmosphere (both naturally and anthropogenically) undergo complex chemical reactions (often driven by solar ultraviolet radiation) that are only just now beginning to be understood. It is not the purpose of this book to expound on this topic; many excellent texts and review articles are available (see the Bibliography) that the reader is urged to consult. For the present purpose we shall simply tabulate the primary constituents of the atmosphere and indicate those that are known or suspected to be changing.

3.2.1. Atmospheric Composition

Some 99.9% of the Earth's permanent atmosphere is composed of nitrogen (78.09%), oxygen (20.95%), and argon (0.94%). The only other species present in substantial amounts is water vapor, whose relative abundance is wildly variable between essentially zero and an all-time record high of 5%, with an average of about 0.7% (all by volume). The small fractional volume remaining (0.03% of the dry atmosphere) encompasses all other species, the bulk of which, however, constitute the major radiatively active gases and all those involved in the problems outlined above (see Table 3.1).

In addition, the atmosphere contains trace amounts of peroxides, other sulfur and halogen compounds, and assorted higher hydrocarbons. A common factor among these molecules (including, *a fortiori*, water vapor) is that they all have discernible infrared signatures and (by no means coincidentally) many of them are implicated in the "greenhouse" effect.

3.2.2. Compositional Profiles

Most of the important gases are nonuniformly mixed through the atmosphere. They vary (sometimes strongly) in place, time, and altitude. Determination of these *compositional profiles*, as they are called, is a key element in attempting to understand the complex physical and chemical processes in the atmosphere.

Determination of horizontal and temporal variability is a matter of developing a proper sampling strategy, since no sensor can observe everywhere all

Table 3.1. Radiatively Active Minor Constituents of the Atmosphere

Carbon dioxide (CO_2)	344 ppmV (increasing ~ 1 ppmV/year)
Ozone (O_3)	8200 ppbV, 25–30 km midlatitudes (decreasing)
Ozone (O_3)	0–200 ppbV, 0–10 km (increasing)
Carbon monoxide (CO)	5–10,000 ppbV, 0–10 km
Methane (CH_4)	1680 ppbV (increasing $\sim 1\%$ per year)
Nitrous oxide (N_2O)	310 ppbV (increasing)
Nitric oxide (NO)	5–100,000 pptV, 0–10 km
Nitric acid (HNO_3)	100–1000 pptV, 0–10 km
Nitrogen dioxide (NO_2)	5–100,000 pptV, 0–10 km
Ammonia (NH_3)	1–10,000 pptV, 0–10 km
Sulfur dioxide (SO_2)	5–1000 pptV, 0–10 km
CFC 12 (CF_2Cl_2)	360 pptV (increasing)
CFC 11 ($CFCl_3$)	200 pptV (increasing)

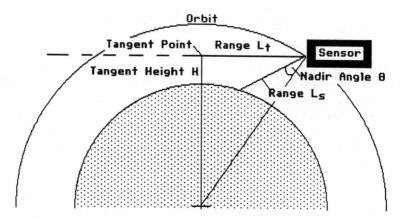

Figure 3.2. Geometry of limb and nadir remote sounding of the atmosphere.

the time. Height variability is a different issue. There are two basic approaches: *limb sounding* and *(near-)nadir sounding*.

Limb sounding (Fig. 3.2) obtains height profiles by looking sideways through the atmosphere. The height resolution is a matter of detailed sensor geometry, but 1–3 km is normal. The defect of the method is that the horizontal resolution in the line of sight is not very good (100–200 km). On the other hand, the long path length makes trace species easier to detect, so the technique is widely used in stratospheric remote sensing for such problems as ozone depletion.

Near-nadir sounding provides much better horizontal resolution at the expense of poorer vertical resolution. As we shall see in the following section, typical height resolutions are 0.5–1.0 scale height (i.e., 4–8 km). The choice between the two methods in the troposphere is quite unclear, so (as we shall see in Chapter 5) at least one planned tropospheric remote sensor will use both approaches.

3.3. RADIATIVE TRANSFER

Radiative transfer is a theory that permits the computation of the signal received at the input to the remote sensor in terms of a parameter called *radiance* (more properly *spectral radiance*, although "spectral" is often left to be understood). The units are (for our purpose) $W \cdot cm^{-2} \cdot sr^{-1} \cdot (cm^{-1})^{-1}$. Although radiative transfer theory can become very complex and is the subject of numerous textbooks (see the Bibliography), the underlying idea is quite simple. In words, it can be stated as "the radiance received at some plane B equals the radiance emitted at another plane A plus whatever came before, and the whole multiplied by the transmittance between A and B." So far, this is all rather obvious. We can began to build up a mathematical description by reference to Fig. 3.3, which divides up the atmosphere into layers (strictly speaking, concentric shells), ignoring clouds and scattering. Within each layer, the temperature, pressure, and density are assumed to be constant.

When LTE holds (as it does anywhere in the Earth's atmosphere up to altitudes of 80 km or more), the radiance J_e emitted by any surface or layer is the product of its *emittance* $\varepsilon(\bar{v})$ and the Planck function $B(\bar{v}, T)$:

$$J_e(\bar{v}) = \varepsilon(\bar{v}) \cdot B(\bar{v}, T) = \varepsilon(\bar{v}) \cdot c_1 \bar{v}^3 / [\exp(c_2 \bar{v}/T) - 1] \qquad (3.5)$$

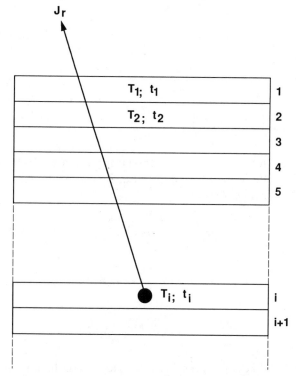

Figure 3.3. The basic tenets of radiative transfer. $T =$ temperature; $t =$ transmittance.

where $c_1 = 1.1911 \times 10^{-12}$ for $B(\bar{v}, T)$ in $W \cdot cm^{-2} \cdot sr^{-1} \cdot (cm^{-1})^{-1}$ and $c_2 = 1.4388$. Note that ε is an intrinsic property of the emitting material (gas, liquid, or solid), whereas B depends only on the material's temperature (and both on frequency, of course). Since gases are nonreflective, it also follows from Kirchhoff's law that $\varepsilon(\bar{v}) = 1 - t(\bar{v})$, where t is the transmittance. Thus the received radiance J_r from the ith and all higher layers in Fig. 3.3 is (leaving the dependence on $\bar{v}$ implicit for the moment)

$$J_r = (1 - t_1) \cdot B_1(T_1) + (1 - t_2) \cdot B_2(T_2) \cdot t_1 + (1 - t_3) \cdot B_3(T_3) \cdot t_1 \cdot t_2 + \cdots$$
$$+ (1 - t_i) \cdot B_1(T_i) \prod_1^{i-1} t_j$$

or

$$J_r = B_1(T_1) + B_2(T_2) \cdot t_1 + B_3(T_3) \cdot t_1 \cdot t_2 + \cdots + B_i(T_i) \prod_1^{i-1} t_j$$
$$- \left[B_1(T_1) \cdot t_1 + B_2(T_2) \cdot t_1 \cdot t_2 + B_3(T_3) \cdot t_1 \cdot t_2 \cdot t_3 + \cdots + B_i(T_i) \prod_1^{i} t_j \right]$$
$$= \sum_j B_j(T_j) \cdot \prod_{j-1} t_k - \sum_j B_j(T_j) \cdot \prod_j t_k \tag{3.6}$$
$$\approx \sum_j B_j(T_j) \left(\prod_{j-1} t_k - \prod_j t_k \right)$$

Now, the product terms are simply the total transmittance τ from the layer to space. If the altitude of the ith layer is z, Eq. 3.6 can be directly generalized (reinserting the dependence on $\bar{v}$)

$$J_r(\bar{v}, z) = \int_{\tau(z)}^{1} B(\bar{v}, T(z)) \, d\tau \tag{3.7}$$

It is convenient to change the integration variable to z:

$$J_r(\bar{v}, z) = \int_{z}^{\infty} B(\bar{v}, T(z)) \frac{\partial \tau}{\partial z} \, dz \tag{3.8}$$

Equation 3.8 is the key to atmospheric remote sensing because it can have some interesting properties (and is also the reason for spending so much effort to derive it). Figure 3.4a shows a sketch of how atmospheric transmittance varies with altitude (at some chosen frequency). A photon emitted low down in the atmosphere has very little chance of reaching space: the transmittance is 0. Near the top of the atmosphere, the chance is very high: the transmittance approaches 1. Evidently, at some intermediate altitude we must go through a transition from low to high transmittance, as shown. Now the kernel of Eq. 3.8 is the height derivative of the transmittance (shown in Fig. 3.4b). Observe that it is strongly peaked at a particular altitude, forming a *weighting function* for the integral. The consequence is that the integral is negligible except over a narrow range of altitudes and provides a means of mapping spectral frequency to altitude because weighting functions for different frequencies peak at different altitudes. Note, however, that not all frequencies generate "good" (i.e., single-valued) weighting functions, and one of the major tasks of the designer of an atmospheric remote sensor is to discover the correct frequencies for the system and measurement goals chosen.

3.3.1. Radiative Transfer in a Clear Atmosphere

For a downlooking remote sensor, the foregoing arguments are easily extended to add in the contributions of surface reflected sunlight; surface emission; and a term to account for downwelling atmospheric emission reflected back upward from the surface. In order to show the influence of the sensor on the signal, the result should be convolved with the absolute ILS. For simplicity, however, we assume a rectangular ILS so the convolution reduces to a simple integration over the spectral resolution, as shown in Eq. 3.9:

$$J(\Omega, \bar{v}) = \int_{\bar{v}-\delta\bar{v}/2}^{\bar{v}+\delta\bar{v}/2} \Phi(\bar{v}, \bar{v}') \left[\left\{ A(\bar{v}') \cdot F(\Omega, -\Omega_0, \bar{v}') \cdot E_s(\bar{v}') \cdot \Omega_s \cdot \tau(-\Omega_0, \infty, z_0, \bar{v}') \right. \right.$$

$$\text{(Instrument)} \quad \text{(Reflected sunlight term)}$$

$$+ \varepsilon(\Omega, \bar{v}') \cdot B(\bar{v}', T_b)$$

$$\text{(Surface emission term)} \tag{3.9}$$

$$+ A(\bar{v}') \int_{2\pi} F(\Omega, -\Omega_0, \bar{v}') \int_{\infty}^{z_0} B(\bar{v}', T(z)) \frac{\partial \tau(-\Omega', z, z_0, \bar{v}')}{\partial z} dz d\Omega' \right\}$$

$$\cdot \tau(\Omega, z_0, \infty, \bar{v}')$$

$$\text{(Downwelling, back-reflected, atmospheric emission term)}$$

$$+ \int_{z_0}^{\infty} B(\bar{v}', T(z)) \frac{\partial \tau(-\Omega', z, z_0, \bar{v}')}{\partial z} dz \right] d\bar{v}'$$

$$\text{(Upwelling atmospheric emission term)}$$

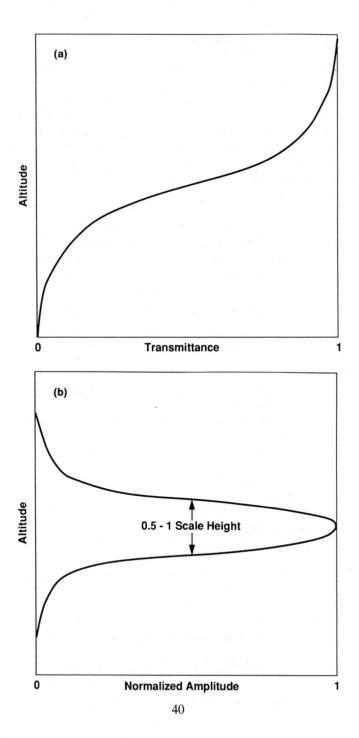

where

$J(\Omega, \bar{v})$ = radiance at frequency $\bar{v}$ into upward, directed, solid angle Ω

$\delta\bar{v}$ = effective spectral resolution

$\Phi(\bar{v}, \bar{v}')$ = instrumental absolute ILS (impulse response)

$\bar{v}'$ = frequency integration variable

$A(\bar{v}')$ = surface albedo

$F(\Omega, -\Omega_0, \bar{v}')$ = surface biconical reflectance function for incident (downward) solid angle $-\Omega_0$ and emergent (upward) solid angle Ω

$= (\cos\theta)/\pi$ for a Lambertian surface, where θ is the zenith angle associated with Ω

$E_s(\bar{v}')$ = disk-average solar radiance

Ω_s = solar solid angle at Earth (6.8×10^{-5} sr)

$\tau(\Omega, z, z', \bar{v}')$ = atmospheric transmittance at frequency $\bar{v}'$ in a direction Ω between altitudes z and z'

$\varepsilon(\Omega, \bar{v}')$ = surface emittance at frequency $\bar{v}'$ in the upward direction Ω

$B(\bar{v}', T_b)$ = Planck function for surface temperature T_b

$B(\bar{v}', T(z))$ = Planck function for atmospheric temperature $T(z)$ at altitude z

$\partial\tau(\Omega, z, z', \bar{v}')/\partial z$ = atmospheric weighting function for direction Ω

The product of the weighting function and the Planck function (i.e., the integrands on the last two lines of Eq. 3.9) is called the *contribution function* since it expresses the contribution of a particular altitude z to the total outgoing radiance.

The transmissions are a product over all N species present in the line of sight:

$$\tau(\theta, z, \bar{v}) = \prod_{i=1}^{N} \exp\left[-\int_z^\infty L(\theta, z') \cdot m_i(z') \cdot \kappa_i\{\bar{v}', z', T(z')\} \, dz' \right] \quad (3.10)$$

where $L(\theta, z)$ is the ray path length [$\propto \sec(\theta)$ in a plane-parallel, unrefracted

Figure 3.4. (a) Sketch of the transmittance to space from a strong atmospheric absorber. (b) The height derivative (weighting function) of Fig. 3.4a.

geometry) and m_i is the volume mixing ratio of the ith species with absorption coefficient κ_i.

3.3.2. Discussion

The activity called *remote sensing* attempts to invert Eq. 3.9: given a set of radiances vs. frequency (otherwise known as a spectrum) what are—

   m_i, the mixing ratios of all N species as a function of z;

   $T(z)$, the atmospheric temperature profile;

   T_b, the surface sensible temperature;

   $\varepsilon(\Omega, \bar{\nu})$, the surface emittance function;

   $A(\bar{\nu})$, the surface albedo;

   $F(\Omega, -\Omega_0, \bar{\nu}')$, the surface biconical reflectance function?

A basic premise is to select frequencies (and times of day) so that, as far as possible, one of the terms in the equation dominates. While this simplifies the analysis, it does not render it trivial! Developing good retrieval algorithms is a significant research topic that is far from solution.

The expressions were developed for a downlooking sensor, but the same equations hold for limb emission sounding if the first three (surface-related) terms are omitted. While the equation, therefore, seems much simpler, the geometry of the light path (especially in the lower atmosphere) becomes much more complicated because of strong refraction effects.

An alternative—and popular—approach is *solar occultation*, in which the sun is used as a very bright extra-atmospheric light source. Radiative transfer for this case devolves virtually to a simple transmission measurement (i.e., *all* emission terms can be ignored), especially—as is possible from space—if the sun can be observed well above the atmosphere (above 200 km, say) and the atmospheric data ratioed against the pure solar spectrum. The defects of the method are that (a) the measurements are necessarily limited to the local sunrise or sunset times, eliminating the possibility of observing diurnal varia- tions, and (b) for a given orbit or location the latitudes sampled vary only slowly with time. Global coverage in a reasonable time is therefore impossible. In an effort to overcome these limitations, some investigators are examining the possibility of using the moon and/or stars for the same purpose. While the instrumental constraints that this imposes are formidable (see Chapter 4), it is worth noting that the technique of stellar occultation has been used by

──▶

Figure 3.5. An overall view of the vertical transmission of the Earth's atmosphere. (Courtesy U.S. Air Force Geophysical Laboratory.)

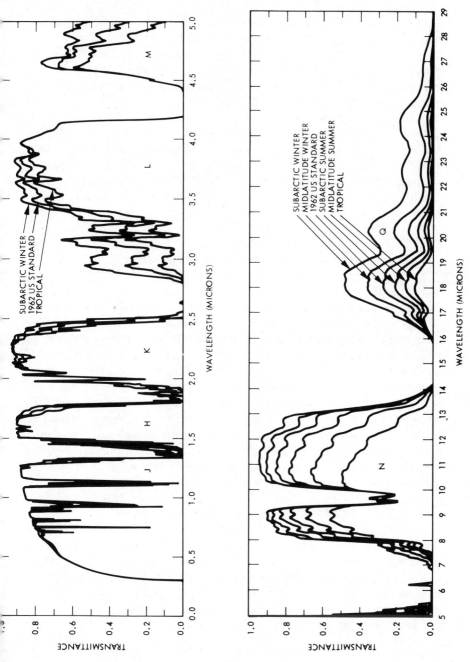

43

planetary astronomers for years not only for atmospheric studies but also to map faint rings around the outer planets.

Figure 3.5 shows an overall view of the vertical transmission of the atmosphere from space to the surface from the visible to the mid-infrared. The letters in the (more or less) transparent regions ("atmospheric windows") are used by astronomers for identification purposes.

In the real world, we may not be able to ignore scattering, especially at visible and near-infrared wavelengths. This introduces enormous complexity into radiative transfer because (a) the photons do not travel directly from source to sensor and (b) some of the photons reaching the sensor will not have originated from the required target. Indeed, at wavelengths below about $0.5 \, \mu m$, *most* of the photons are backscattered sunlight from the atmosphere. Retrieval of surface properties becomes difficult.

3.4.   REMOTE SENSING SPECTROMETRY

When a molecule absorbs or emits a photon, the energy state necessarily changes. In general, the energy changes can occur in *translation* (heating or cooling) or as a change in *electronic, vibrational,* or *rotational* state. The Born–Oppenheimer approximation asserts that, because the energies involved in these changes are dramatically different, they can be treated independently. Much of the time, this is true. For solids and liquids, of course, molecular rotation and translation are impossible, so they respond with changes in electronic and vibrational state. Except for translation (which permits an essentially continuous range of energies), the states of the molecule are discrete (quantized). Consequently, emission and absorption of light can occur only at well-defined frequencies. Furthermore, the frequency set for each molecular species is unique, providing a "fingerprint" for that species. Even isotopic substitutions produce recognizable differences in the spectrum.

3.4.1.   The Mechanistic Molecular Model

One can achieve surprisingly good insight into molecular spectra with a simple mechanical model of a molecule—a set of point masses connected together with weightless springs. While such a molecule lacks an electronic cloud to generate electronic spectra, it transpires that the vast majority of infrared spectra are concerned only with vibration and rotation. In general, molecules remain in their electronic ground state unless excited by visible or ultraviolet photons.

Molecular rotation occurs around the three principal moments of inertia. The energy associated with changes in rotational state are typically in the range 0.1–$10.0\,\mathrm{cm}^{-1}$. [*Note*: To simplify calculations, the usual units of energy (joules, J) are divided by Planck's constant and the velocity of light—giving cm^{-1}—so that emitted and absorbed frequencies are simply the differences between the values assigned to each level.] The mechanistic model also demonstrates *centrifugal stretching*: as the molecule "spins" at greater and greater rates (higher energy), the springs "stretch" and produce an observable change in the spacing between successive energy levels.

A molecule composed of $N(N \geqslant 3)$ atoms can vibrate in $(3N - 6)$ modes, although some of the modes may be degenerate (identical). Linear molecules (including all diatomics such as CO) have $(3N - 5)$ modes. Typical vibrational energy changes lie in the range 100–$4000\,\mathrm{cm}^{-1}$. Since a molecule can vibrate and rotate simultaneously, one typically observes a *vibration-rotation band*—a series of discrete rotational transitions centered on a particular vibrational frequency. A simple example (CO) is shown in Fig. 3.6.

The simple model breaks down in one important regard: the vibrational motion in reality is somewhat anharmonic, which permits overtones and

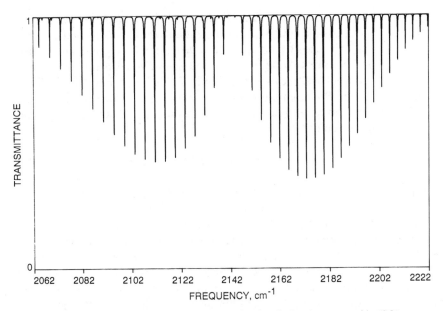

Figure 3.6. The fundamental vibration–rotation band of carbon monoxide (CO).

combinations of the fundamental vibrational frequencies to become active. Since transitions between any of these vibrational states are usually allowed (in the quantum mechanical sense), spectra can become quite complex—with the complexity generally increasing with increasing temperature.

Another major source of complexity lies in the fact that the three principal moments of inertia may be identical [called a *spherical rotor*, methane (CH_4) being an example]; two may be identical, giving rise to a *symmetric rotor* [ammonia (NH_3) is an example, as are (with a slight extension of the definition) all diatomic and linear molecules such as carbon dioxide (CO_2)]; or all three may be different, resulting in an *asymmetric rotor* [water vapor (H_2O), for instance]. Generally speaking, asymmetric molecules have the most complex and irregular spectra (virtually random) and symmetric molecules the simplest and most regular. In any event, it transpires that the energy states of molecules (and therefore their spectra) are, by and large, not calculable from first principles. They must be determined empirically through laboratory investigations. These tend to be extremely time consuming; it is by no means unusual for individuals to spend a lifetime on a single molecule!

3.4.2. Intensities

The mechanistic model provides an understanding of *what* transitions (i.e., frequencies) are possible but gives little insight into the *likelihood* of the transition occuring. This leads to a discussion of *line intensities* (sometimes called *line strengths*). There are two major factors: the *transition probability*, a quantum mechanical concept well beyond the scope of this book (but see the Bibliography), and the Boltzmann distribution. Elementary statistical mechanics shows that the number of molecules N having an energy E above the ground state ($E = 0$) is given by

$$N = N_0 \exp\left[-E/(kT)\right] \tag{3.11}$$

where N_0 is the total number of molecules, k is Boltzmann's constant (0.695 for E in cm^{-1}), and T is the absolute temperature. This is a very steeply falling function. For example, the fraction of molecules having an energy of $1000\,cm^{-1}$ at room temperature is less than 1%. On the other hand, this same fact is part of the reason that atmospheric temperatures can be determined by remote sensing: atmospheric spectra are *very* temperature sensitive. Again, intensities are not accurately calculable from first principles and must be measured (under very carefully controlled conditions) in the laboratory. Provided that the energy levels are known, however, theory does permit a prediction of how

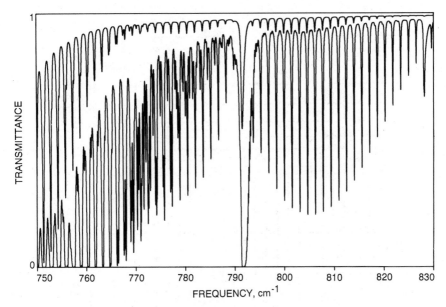

Figure 3.7. Example of the temperature dependence of a rotation–vibration band of carbon dioxide (CO_2). The gas abundance is the same in both cases; only the temperatures differ: *upper*, 200 K; *lower*, 300 K. Note also the strong Q branch in the middle of the figure.

the intensity will vary with temperature. This capability is a key factor in remote sensing. An example is shown in Fig. 3.7.

3.4.3.   Terminology

For historical reasons, the rotational structure on the low-frequency side of a band center is called a *P branch* and other that on the high-frequency side an *R branch*. At the band center (which corresponds to the frequency of the change in vibrational energy), many (but not all) bands will also show a closely packed group of lines called a *Q branch* (the strong feature in the middle of Fig. 3.7): *P* and *R* branches result from a 1 unit change in rotational quantum number; a *Q* branch from no change (Fig. 3.8).

3.4.4.   Line Shapes

The "discrete" energy levels of a molecule (Fig. 3.8) are not quite identical from molecule to molecule of the same species. For remote sensing, the two important

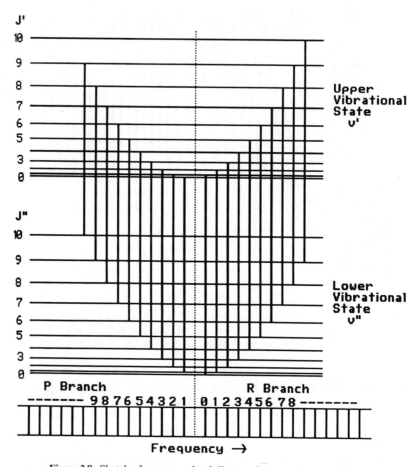

Figure 3.8. Sketch of an energy-level diagram for two vibrational states.

sources of this effect are the Doppler effect due to the translational motion of the molecules and distortions induced by the presence of other molecules. Not surprisingly, then, the composite line broadening due to the Doppler effect is temperature dependent, whereas the broadening due to other molecules (often called Lorentz broadening) is a function of both pressure and temperature. In the Earth's atmosphere, Doppler broadening dominates above about 40 km; Lorentz broadening below 10 km. The "difficult" region lies in between, when both matter.

The absorption coefficient for a Doppler line is

$$\kappa_D(\bar{v} - \bar{v}_0) = \frac{S(\bar{v}_0)\exp\left[-\ln 2 \cdot \{(\bar{v} - \bar{v}_0)/\alpha_D\}^2\right]}{\alpha_D[(\ln 2 \cdot \pi)^{1/2}]} \tag{3.12}$$

where $S(\bar{v}_0)$ is the line strength and α_D, the Doppler width, is given by

$$\alpha_D = (\bar{v}_0/c)[(2 \cdot \ln 2 \cdot kT/m)^{1/2}] = 3.581 \times 10^{-7}\bar{v}_0[(T/m)^{1/2}] \tag{3.13}$$

for m in atomic mass units.

The absorption coefficient for a Lorentz line is

$$\kappa_L(\bar{v} - \bar{v}_0) = \frac{S(\bar{v}_0) \cdot \alpha_L}{\pi[(\bar{v} - \bar{v}_0)^2 + \alpha_L^2]} \tag{3.14}$$

where α_L, the Lorentz width, is directly proportional to pressure and roughly inversely proportional to $T^{1/2}$.

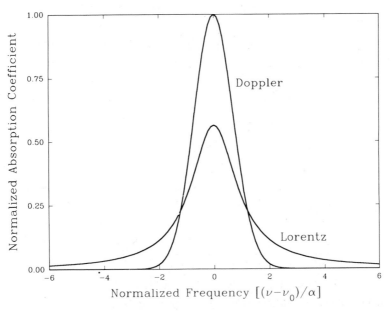

Figure 3.9. Normalized shapes of Doppler and Lorentz lines.

When both Doppler and Lorentz broadening are important, the two functions must be convolved (giving rise to the *Voigt* function) to determine the absorption coefficient. In any event, the transmittance τ is

$$\tau = \exp(-\kappa L) \qquad (3.15)$$

where L is the column density. Figure 3.9 shows normalized versions of the two basic line shapes.

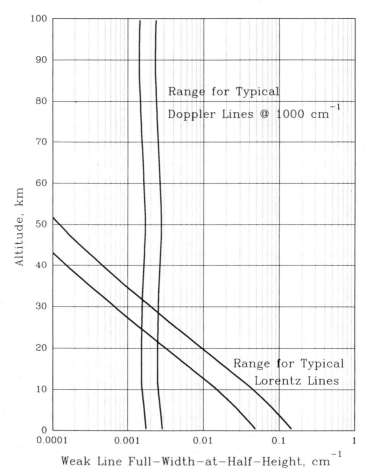

Figure 3.10. Altitude variation of typical weak Lorentz- and Doppler-broadened lines through the Earth's atmosphere (cf. Fig. 3.1).

It is interesting that, on a frequency scale, the Lorentz widths ($= 2\alpha_L$) of most molecules fall near $0.1 \, cm^{-1}$ at 1 atm pressure (within a factor of 2) *independent of frequency*. Thus, as an asymptotic expansion of Eq. 3.15 for small κ shows, "weak" spectral lines (typically less than 30% central depth) will display a similar width. Figure 3.10 shows how such a "typical" weak Lorentz line will vary with altitude in the atmosphere, together with representative Doppler widths at $1000 \, cm^{-1}$ for water vapor ($m = 18$) and carbon dioxide ($m = 48$). The crossover in the lower stratosphere is clearly evident by comparing Fig. 3.10 with Fig. 3.1. The crossover altitude for any other frequency can be estimated by mentally shifting the Doppler range left or right in the figure. For example, at $500 \, cm^{-1}$ ($20 \, \mu m$) it occurs near 30 km, whereas at $5000 \, cm^{-1}$ ($2 \, \mu m$) it is nearer 15 km. As a rule of thumb, if either one is ten times the other, the smaller can be ignored. Otherwise, the Voigt function must be used. Since its direct calculation is extremely time consuming, numerous authors have published lookup tables to simplify the computation. One of the best is that of Humlicek (1982).

3.4.5. Quantitative Atmospheric Spectral Analysis

The combination of the foregoing themes of atmospheric physics and chemistry, radiative transfer, and molecular spectrometry has enabled a new field of *quantitative* atmospheric remote sensing to emerge. Current and planned remote sensors not only *identify* atmospheric constituents but *quantify* (in three spatial dimensions and in time) their distribution. Application of these methods permits atmospheric temperatures to be determined to 1–2 K on a routine basis and compositions to be measured to 1% in favorable cases (Rodgers, 1976).

The best way of illustrating this assertion is through an example. Figure 3.11 shows an actual atmospheric spectrum acquired with the Atmospheric Trace Molecule Spectrometer (ATMOS) FTS, together with a computer-generated model. Were it not for the random noise on the field data, it would require an expert eye to detect any difference. Bear in mind that the process of creating the model spectrum generates a chemical and physical data base that contains *all* the information extractable from the field data.

In Chapter 2 we discussed spectral resolution in abstract terms without specifying on what grounds a choice is to be made. It is demonstrable (see the Appendix) that the optimum information transfer occurs when the spectral resolution matches the width of the spectral features. While this might seem to be intuitively obvious, in fact throughout much of the literature statements such as "low resolution offers better signal-to-noise, therefore superior results

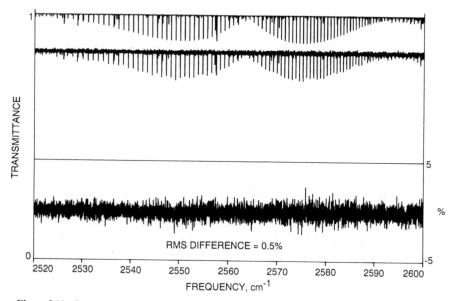

Figure 3.11. Comparison of an actual ATMOS spectrum of N_2O acquired at 41 km tangent height and a computer-generated model: *upper curve*, model; *middle curve*, ATMOS data; *lower curve*, difference on an expanded scale.

emerge" and, at the opposite extreme, "resolution much finer than the intrinsic line width is essential to eliminate distortion" are commonly found. While it is certainly true that the information content is not sharply peaked at the matching condition, it is nevertheless true that all high-quality, functional, remote sensing FTS adhere closely to this principle. Thus systems optimized for the troposphere commonly have a spectral resolution near 0.1 cm^{-1}, whereas stratospheric systems usually achieve 0.01 cm^{-1} or better.

3.4.6. The Spectra of Solids and Liquids

Since rotation is inhibited in solids and liquids, the rotational substructure of a gaseous molecule is absent. Furthermore, the energy levels of solids and liquids are much "fuzzier." Consequently, their spectra tend to be broad and relatively featureless (by comparison to gases). Another factor is that natural materials rarely show much transparency—the interaction of radiation with the material takes place within a few wavelengths of the surface. There are several consequences:

- The spectra are strongly affected by the physical state of the material—smooth, solid, finely divided, etc.—but no information can be directly gleaned about vertical extent.

- The spectra are almost always from an inhomogeneous mixture of many materials. This "mixed pixel" problem is one of great importance in the interpretation of surface remote sensing data.

- The spectral features are often generated not by the bulk properties of the material but by the presence of minute impurities in the lattice. A dramatic example of this is afforded by aluminum oxide (corundum). In

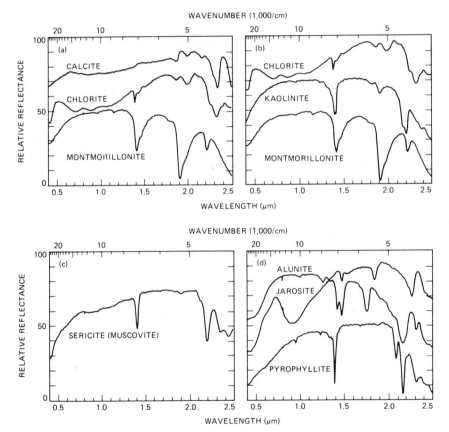

Figure 3.12. Examples of laboratory reflectance spectra of common minerals. Note the substantial wavelength/frequency scale differences between these and atmospheric spectra.

its pure crystalline state it is water white and totally transparent from the ultraviolet out to $5\,\mu$m in the infrared. The addition of a trace of chromium oxide converts it into the brilliantly colored red gemstone ruby. However, if iron and titanium oxides are substituted at similar concentrations, one gets blue sapphire instead! The spectral characteristics (the colors) are due to the impurities, not the host material.

- No general theory of band shape and strength exists for solids and liquids. Remote sensing analysing must therefore depend on empirical comparisons to laboratory samples and *in situ* field data.

Figure 3.12 shows some examples of solid surface spectra. Compare the frequency/wavelength scale to that in Fig. 3.11.

3.4.7. Discussion

Perhaps the major difference between surface and atmospheric remote sensing lies in the spatial resolution requirements. Since atmospheres are motile, their properties tend to be nonlocalized and spatial resolutions of hundreds of meters to kilometers suffice unless specific issues such as plumes from factories are the target. By contrast, surface remote sensors are generally intended for high-resolution mapping. Even commercial systems such as SPOT achieve 10-m resolution, and 5-m systems are in development. The requirements are obviously based on the scale over which the signatures are expected to change.

CHAPTER

4

REAL FOURIER TRANSFORM SPECTROMETERS

In this chapter we shall investigate some of the properties required to convert the Michelson interferometer into a usable FTS.

We begin with the important consideration of how to compute the expected signal-to-noise ratio for an FTS, using a novel approach that is significantly different from that used in existing textbooks but which makes the contribution of each subsystem much easier to recognize (and therefore control).

The next topic is sampling. The proper sampling of an interferogram is absolutely crucial to the success of an FTS and generates the most stringent requirements.

Following this, we consider some widely used optical configurations for remote sensing systems and their relative advantages and disadvantages.

We conclude with a look at some potential problem areas, considering their avoidance and/or solution.

4.1. THE ESTIMATION OF SIGNAL-TO-NOISE RATIO

4.1.1. Sources and Backgrounds

Since the sources investigated by remote sensing are immutable, it follows that the only way to achieve a satisfactory signal-to-noise ratio (SNR) is to minimize the sources of noise. In many (if not most) modern remote sensors, the most serious noise source is the background. Backgrounds corrupt signals in two ways. First, they can create a contrast problem because the wanted signal "sits" on a nonzero level; as an example, the background generated by atmospheric emission in infrared astronomical investigations can be orders of magnitude greater than the signal from the star or planet. Second, the process of photon emission is statistical. To a very good approximation, if one expects to receive N photons from the sum of the source and the background in a given time interval, the standard deviation will be $N^{1/2}$; this is a *definition* of noise.

The expected signal can be computed by solving the equation of radiative

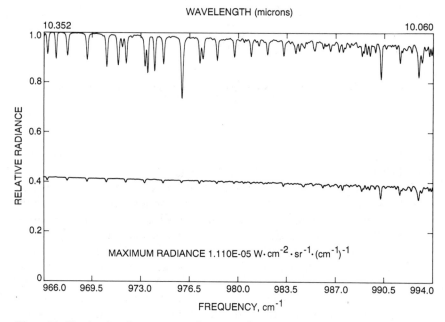

Figure 4.1. Simulated nadir spectra seen from space for two different surface temperatures and atmospheric conditions. Principal absorbers are CO_2, H_2O, and O_3.

transfer (Eq. 3.9) for the range of expected conditions. While this is time consuming, it provides the best estimate of what one expects to see; a couple of examples of this approach are shown as Figs. 4.1 and 4.2. An alternative and widely used method is to parameterize the problem by *brightness temperature*.

Imagine a source whose sensible temperature is T_s with an emittance ε. The brightness temperature T_b is defined by the Planck function:

$$B(\bar{v}, T_b(\bar{v})) = \varepsilon(\bar{v}) \cdot B(\bar{v}, T_s) \tag{4.1a}$$

or, more explicitly,

$$T_b(\bar{v}) = c_2 \bar{v} / \ln[1 + \{\exp(c_2 \bar{v}/T_s) - 1)\}/\varepsilon(\bar{v})] \tag{4.1b}$$

where $c_2 = 1.4388$. Note that

     unless $\varepsilon \equiv 1$, $T_b < T_s$;

T_b is frequency dependent.

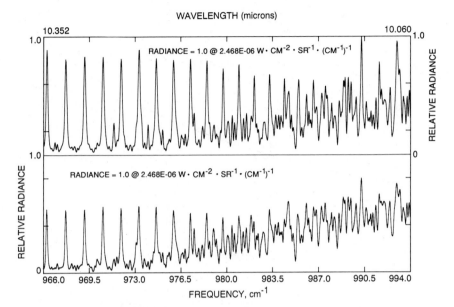

Figure 4.2. The same spectral region and atmospheric conditions as Fig. 4.1 as seen in emission at the limb (tangent height 10 km). Spectra offset one-half frame for clarity, but the vertical radiance scale is the same for both.

Surface thermal remote sensors, for example, measure T_b, not the true temperature T_s. While it is true that many natural surfaces have $\varepsilon \to 1$, the quite common practice of citing T_b as though it had some physical significance is to be deplored. Brightness temperature is a useful tool, no more.

While (as already noted) the source inherently contributes noise owing to the statistical nature of photon emission, *backgrounds* most frequently originate within the sensor and its immediate environment. The definition is inevitably loose because one observer's background can be another's signal. A case in point is thermal remote sensing of the Earth. To the surface scientist, the atmosphere generates a "background." For the atmospheric scientist, the reverse is true: it is the surface that creates the background.

The two major sources of instrumental background are thermal emission and stray light. Stray light is notoriously difficult to control, and the shorter the wavelength, the worse it gets. The classical approach has been to paint all nonoptical surfaces black. Unfortunately, many "black" surfaces are black only at visible wavelengths. An example is black-anodized aluminum. At wave-

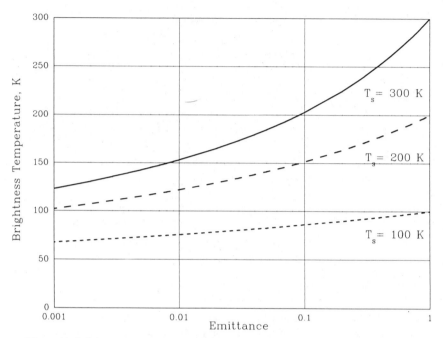

Figure 4.3. Brightness temperature as a function of sensible temperature and emittance.

lengths beyond about 1 μm, black anodizing actually *enhances* reflectance over the untreated material! Proper baffling is a better solution. /

Instrumental thermal emission is a fact of life. The only ways of controlling it are (a) to cool the system and (b) to keep the effective emittance as low as possible. The first option may not be possible because of factors such as the environment in which the instrument must function, not to mention cost and lifetime. The second is cheaper. Figure 4.3 shows how system brightness temperature declines with reduced emittance. Highly polished surfaces such as mirrors should have a negligible emittance, but mountings and the enclosures surrounding them radiate very effectively (indeed, virtually as a black body), so only a small fraction need be scattered into the optical path for the *system* emittance to be significant. While good data are hard to find in the literature, / it seems that reducing system emittance below 0.1–0.2 is extremely difficult. /

4.1.2. Detectors

Detectors are a topic unto themselves, and it is well beyond the scope of this book to go into detail about the physics of the detection process (some suggested reading can be found in the Bibliography). Furthermore, information on specific detectors is often proprietary; nevertheless, some general comments are in order.

As a rule of thumb, the longer the wavelength one wishes to sense (at least up to region where microwave technology takes over), the colder the detector must be. Through the visible and out to about 1 μm, detector element temperatures rarely need to be below 200 K. From 1 to 17 μm, temperatures are typically in the range 50–80 K. Longer wavelengths usually demand temperatures below 20 K. The basic physics behind these requirements is quite straightforward: in order for a photon of frequency $\bar{v}$ to "liberate" an electron (or a "hole") from its bound condition within the detector (thus making it available to generate a measurable charge, voltage, or current), the effective binding energy must be less than $hc\bar{v}$. However, lattice vibrations (phonons) carrying energy $\sim kT$ can be equally effective. Thus, in order to avoid thermal noise, one generally requires $T < hc\bar{v}_{min}/(10\ K)$ ($\approx \bar{v}_{min}/10$), where $\bar{v}_{min}$ is the lowest frequency to be detected. Note that this requirement is not simply to achieve a good SNR; many detectors will be physically destroyed if an attempt is made to use them above their design temperatures.

From the ultraviolet to the near-infrared (1 μm), silicon photodiodes and photoemissive devices such as photomultipliers have long held sway. While virtually every user of detectors feels that what is available is never good enough, it does not seem likely that these technologies will be superseded in the near future.

Between 1 and 12 μm, two technologies dominate: indium antimonide (InSb, to 5.5 μm) and mercury cadmium telluride (HgCdTe, or MCT); both operating in a photovoltaic mode (i.e., essentially no current is drawn). InSb is considered a "mature" technology; MCT is still problematic because it is difficult to control material properties.

Between 12 and 35 μm, one must fall back on photoconductors (essentially a variable resistance) through which a bias current must flow. Not only is the current a potential source of noise but the physics of photoconduction is intrinsically noisier (by a factor of $2^{1/2}$) than photovoltaic detection. Consequently, considerable effort is being expended to extend photovoltaic MCT to 17 μm and beyond. Rumors persist of success, but at this writing no such devices are generally available. The detectors most in use in the range 12–30 μm are photoconductive MCT (to 17 μm) and various forms of doped Si (which must, however, be cooled to below 20 K). MCT PC can be used as warm as 80 K (but less is better).

Beyond about 30 μm, the only available technology in common use is the semiconductor bolometer (a temperature-sensitive resistor), although developments such as the blocked-impurity band detector are in progress. Such devices must be cooled to 4.2 K or less.

Most type of detectors are available as one- and two-dimensional arrays, with and without on-chip multiplexers and/or preamps, and the size of the arrays is constantly increasing. Indeed, especially at the shorter wavelengths, the sizes are such as to cause serious distress to existing data systems.

Postdetector signal chains are also widely considered to be a "mature" technology. Nevertheless, they cannot be taken lightly. Most detectors require a preamp that operates at, or near to, the detector temperature with leads that are as short as ingenuity can make them. One problem that an FTS faces (especially with InSb) is that the *frequency response* of many detector systems is quite poor. As we shall see later on, an FTS often requires response to frequencies approaching 100 kHz. This, too, can create problems for the designer.

4.1.3. Radiometric Models

For many years, it has been the practice to characterize infrared detectors by a parameter called *detectivity* (D^*), whose unit is the *jones* (1 jones = 1 cm·Hz$^{1/2}$·W^{-1}, essentially the SNR to be expected from a 1-cm detector in a 1-s integration for a 1-W input). It was a useful parameter when most detectors were limited in performance by internal noise sources; that is no longer the case. Most modern detectors are limited by the Poisson statistics of discrete electron events (commonly known as *counting statistics*) even if no actual counting is performed.

The basic equation is this alternative approach is very simple:

$$\text{SNR} = \frac{\text{number of signal photoelectrons}}{(\text{total number of carriers from all sources})^{1/2}} \tag{4.2}$$

The computation of the numbers to be inserted into Eq. 4.2 is performed with the aid of a *radiometric model*. A detailed radiometric model takes into account the temperature, emittance, transmittance, reflectance, and scattering of every surface in the line of sight from source to detector. That is, every surface both relays the photons originating in front of it and is itself a source of additional (usually unwanted) photons. Within the detector, the photons are converted into photoelectrons with a certain efficiency (the *quantum efficiency*) and further noise electrons are added by the solid-state physics of the detection mechanism and the subsequent electronics.

While such a detailed model is, indeed, necessary for exact calculations, a great deal can be accomplished with a simpler "lumped parameter" model that divides the system into regimes of common temperature and assigns an effective transmittance, emittance, and étendue to each block. Such a model for a generalized remote sensing spectrometer might look like Fig. 4.4.

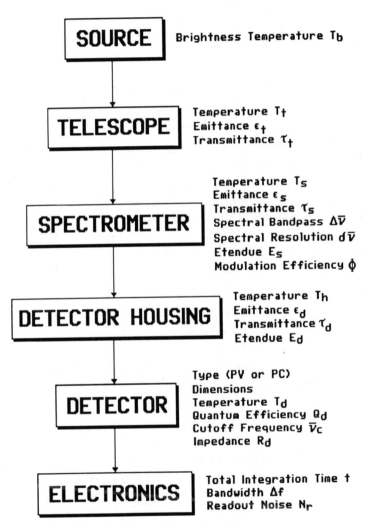

Figure 4.4. A block subsystem radiometric model.

4.1.4.   Fourier Transform Spectrometer

Let

N_1 = number of electrons per second ($e^- \cdot s^{-1}$) generated within the detector and readout system (detector noise)

N_2 = number of $e^- \cdot s^{-1}$ generated by the detector housing internal background

n_3 = number of $e^- \cdot s^{-1} \cdot (cm^{-1})^{-1}$ from instrument body emission

n_4 = number of $e^- \cdot s^{-1} \cdot (cm^{-1})^{-1}$ from pointing and telescope system emission

n_5 = number of $e^- \cdot s^{-1} \cdot (cm^{-1})^{-1}$ from source emission

n_6 = number of $e^- \cdot s^{-1} \cdot (cm^{-1})^{-1}$ from solar reflection

$\Delta\bar{v}$ = width of band-limiting filter (cm^{-1})

$\delta\bar{v}$ = spectral resolution (cm^{-1})

Φ = modulation efficiency

t = total integration time per interferogram

The signal count $S = \delta\bar{v} \cdot \Phi(n_5 + n_6)t\, e^-$ and the noise count $C = (N_1 + N_2)t + \Delta\bar{v} \cdot (n_3 + n_4 + n_5 + n_6)t\, e^-$. Thus the mean SNR (averaged across the band Δv) is

$$\text{SNR} = \frac{\delta\bar{v} \cdot \Phi(n_5 + n_6)t}{[\{(N_1 + N_2) + \Delta\bar{v} \cdot (n_3 + n_4 + n_5 + n_6)\}t]^{1/2}} \tag{4.3}$$

where

$$N_1 = P[4kT_d\,\Delta f/(R_d e)]^{1/2} + N_r^2\,\Delta\bar{v}/(\delta\bar{v} \cdot t) \tag{4.3.1}$$

and

$P = 1$ for a photovoltaic detector; 2 for a photoconductor

k = Boltzmann's constant ($1.38066 \times 10^{-23}\, J \cdot K^{-1}$)

T_d = detector element temperature

Δf = detector system electrical bandwidth

R_d = detector effective resistance

e = charge on the electron ($1.6022 \times 10^{-19}\, C$)

N_r = read noise (applicable only if multiplexed focal planes are used)

$$N_2 = E_d Q_d \int_{\bar{v}_c}^{\infty} B(\bar{v}, T_h)\, d\bar{v} \tag{4.3.2}$$

and

E_d = detector internal étendue $\approx \pi \cdot$(detector area) for a properly cold-stopped detector subsystem

Q_d = detector quantum efficiency

$\bar{v}_c$ = detector low-frequency cutoff

T_h = temperature of the detector cavity

$B(\bar{v}, T)$ = Planck photon function for frequency $\bar{v}$ and temperature $T = 5.9959 \times 10^{10}\, \bar{v}^2/[\exp(1.43883\, \bar{v}/T) - 1]$ photons·s^{-1}·cm^{-2}·sr^{-1}·cm^{-1}

$$n_3 = \varepsilon_s \cdot \bar{B}(\bar{v}, T_s) \cdot E_s \tau_d Q_d \tag{4.3.3}$$

and

ε_s = spectrometer effective emittance

$\bar{B}(\bar{v}, T)$ = band-average Planck photon function

T_s = spectrometer body temperature

E_s = system étendue

τ_d = transmittance of detector system optics and filters

$$n_4 = \varepsilon_t \cdot \bar{B}(\bar{v}, T_t) \cdot E_s \tau_s \tau_d Q_d \tag{4.3.4}$$

and

ε_t = pointing and telescope system effective emittance

T_t = pointing and telescope system temperature

τ_s = transmittance of spectrometer

$$n_5 = \bar{B}(\bar{v}, T_b) \cdot \tau_0 \tau_t E_s \tau_s \tau_d Q_d \tag{4.3.5}$$

and

T_b = source emission brightness temperature

τ_0 = outbound atmospheric transmittance (set = 1 if atmospheric emission is the intended source)

τ_t = transmittance of pointing and telescope system

$$n_6 = \bar{F}(\bar{v}) \cdot G\tau_i \cos(\theta) \cdot \alpha\tau_0\tau_t E_s \tau_s \tau_d Q_d \tag{4.3.6}$$

and

G = sun–Earth geometric factor $\approx 2.165 \times 10^{-5}$ at the mean Earth–sun distance

Table 4.1. A typical Signal-to-Noise Ratio Calculation for an FTS

	Value	Unit
System input parameters:		
Telescope temperature (T_t)	300.0	K
Telescope effective emittance (ε_t)	0.10	
Telescope transmittance (τ_t)	0.930	
Spectroradiometer temperature (T_s)	150.0	K
Spectroradiometer effective emittance (ε_s)	0.32	
Spectroradiometer transmittance (τ_s)	0.340	
FTS modulation efficiency (Φ)	0.90	
System etendue (E_s)	9.45×10^{-5}	cm$^2\cdot$sr
FTS scan speed (OPD)	8.447	cm$\cdot$s^{-1}
Detector type	InSb PV	
Detector element temperature (T_d)	65.0	K
Detector low-frequency cutoff ($\bar{v}_c$)	1800.0	cm^{-1}
Detector housing temperature (T_h)	65.0	K
Detector system transmittance (τ_d)	0.560	
Detector quantum efficiency (Q_d)	0.600	
Detector element size	0.100×0.100	mm
Detector effective resistance (R_d)	1.0×10^8	Ω
Total useful integration time (t)	2.0	s
Spectral resolution (δv)	0.1000	cm^{-1}
Source brightness temperature (T_b)	300.0	K
Solar zenith angle (θ)	180.0	degrees (i.e., midnight)
Surface reflectance (α)	0.150	
Mean outbound atmospheric transmittance (τ_0)	1.000	

Table 4.1 (*Continued*)

	Value	Unit	Noise rank
Expected photoelectron count rates for the band 2050.0–2250.0 cm^{-1}:			
From source emission (n_5)	9.47×10^7	$e^- \cdot s^{-1} \cdot (cm^{-1})^{-1}$	(1)
From telescope (n_4)	1.02×10^7	$e^- \cdot s^{-1} \cdot (cm^{-1})^{-1}$	(2)
From detector (N_1)	3.34×10^6	$e^- \cdot s^{-1}$	(3)
From spectrometer (n_3)	3.51×10^3	$e^- \cdot s^{-1} \cdot (cm^{-1})^{-1}$	(4)
From detector housing (N_2)	8.63×10^{-3}	$e^- \cdot s^{-1}$	(5)
From solar reflection (n_6)	0.00	$e^- \cdot s^{-1} \cdot (cm^{-1})^{-1}$	(6)
Radiances:			
Emission	3.94×10^{-7}	$W \cdot cm^{-2} \cdot sr^{-1} \cdot (cm^{-1})^{-1}$	
Solar reflection	0.00	$W \cdot cm^{-2} \cdot sr^{-1} \cdot (cm^{-1})^{-1}$	
Noise equivalent source radiance (NESR)	4.73×10^{-9}	$W \cdot cm^{-2} \cdot sr^{-1} \cdot (cm^{-1})^{-1}$	
Average spectral signal-to-noise ratio: 83:1			
Interferogram dynamic range (min)	3772	(= 12 bits)	
Detectivity (D^*)	5.31×10^{13}	Jones	
Background flux density	3.50×10^{14}	photons $\cdot s^{-1} \cdot cm^{-2}$	

$\bar{F}(\bar{v})$ = band-average solar radiance

τ_i = inbound atmospheric transmittance

θ = solar zenith angle at the Earth

α = directed surface reflectance

Equation 4.3 warrants examination in some detail. Observe that, under any circumstances, SNR improves only as the square root of the integration time (which equals the *total time to acquire the interferogram*, not the time per sample). Second, if either N_1 or N_2, the detector and electronics terms, dominates the denominator, the SNR becomes independent of the spectral band-pass $\Delta \bar{v}$. This, then, is the classic Fellgett advantage but is also not necessarily a desirable advantage! For preference, one wants *all* noise terms to be as small as possible. However, note that when the detector terms are small, the SNR becomes a function of the total spectral band-pass $\Delta \bar{v}$. In order to improve SNR, we must work with restricted bands. Next, we see that for a given source ($n_5 + n_6$), it becomes very important to keep the instrumental terms ($n_3 + n_4$) small by comparison. This often demands a cooled instrument.

Finally, we see that there is an absolute limit to the SNR determined simply by the available source photons ($n_5 + n_6$).

An example of a typical calculation using this approach is shown in Table 4.1. Note that the expressions for Eq. 4.3 are simple enough that a straightforward interactive computer program is readily devised to permit an investigation of the effect of changing any of the parameters. Table 4.1 was produced using just such a program.

4.1.5.   Dispersive Spectrometer

Using the same notation, we have the equivalent expression for a dispersive spectrometer (DS):

$$\text{SNR} = \frac{\delta\bar{v}\cdot(n_5 + n_6)t/m}{[\{(N_1 + N_2) + \Delta\bar{v}\cdot n_3 + \delta\bar{v}\cdot(n_4 + n_5 + n_6)\}t/m]^{1/2}} \tag{4.4}$$

where m is the number of time slices required to cover the spectral range $\Delta\bar{v}$. For a spectrometer using detector arrays, $m = 1$–100; for a scanning monochromator, $m \approx \Delta\bar{v}/\delta\bar{v} \gg 100$.

Comparing the expressions for the FTS and the DS:

$$\frac{\text{SNR}_{\text{FTS}}}{\text{SNR}_{\text{DS}}} = m^{1/2}\Phi\,\frac{[(N_1 + N_2) + \Delta\bar{v}\cdot n_3 + \delta\bar{v}\cdot(n_4 + n_5 + n_6)]^{1/2}_{\text{DS}}}{[(N_1 + N_2) + \Delta\bar{v}\cdot n_3 + \Delta\bar{v}\cdot(n_4 + n_5 + n_6)]^{1/2}_{\text{FTS}}}$$

Everything being equal, we see that—

- For a spectrometer or detector/electronic-noise-limited system, the FTS gains by a factor of about $\sqrt{m}$.
- For a source or telescope-noise-limited system, the FTS and scanning DS are equivalent.
- An array DS ($m = 1$) appears to be superior by a factor $(\Delta\bar{v}/\delta\bar{v})^{1/2}$.

Everything is *not* equal. Remember that—

- The frequency coverage of an FTS can be very large: 10:1 is routine. Dispersive spectrometers rarely cover more than 2:1.
- The throughput [that is, ($n_5 + n_6$) in the expressions] of an FTS can be orders of magnitude greater than that of a DS.

- The spectral resolution of an FTS is usually much better than that of the DS and is constant across the spectrum.

Thus, for the majority of applications in infrared remote sensing, the FTS is clearly the approach of choice. Nevertheless, there is an "ecological niche" for cooled array spectrometers. It lies in infrared astronomy, where the sources are small (so the greater étendue offered by the FTS is largely ineffective), spectral resolution requirements are often modest, the spectral coverage is controlled by the available atmospheric windows, and the dominant noise source is thermal emission from the atmosphere and telescope. Under these conditions, the cooled array spectrometer can offer better performance than an FTS. Indeed, it is possible to make an intriguing combination of the two: the *cryogenic postdisperser*. In this use, the FTS detector system is replaced by a cooled dispersive array spectrometer of quite modest resolution (5–10 cm^{-1}, say). Each detector now behaves as though it had a very narrow band filter in front of it (thus reducing the FTS instrumental and atmospheric background), but the spectral coverage is ensured by having large numbers of detectors. Experiments with such systems have been made in astronomy, with some success. I am unaware of any usage in other environments.

There is, however, a condition wherein an FTS can actually be *worse* than a dispersive system. We have seen that when noise is independent of the signal, the FTS clearly wins. When noise is proportional to the square root of the signal, the systems are more nearly equivalent. When noise becomes *directly* proportional to the signal, we shall see in Section 4.1.7 that the FTS clearly loses!

4.1.6. The Calculation of D^* and Background Flux Density

The foregoing, unfortunately, takes no account of the fact that detector manufacturers still think in terms of detectivity (D^*) and become dismayed when presented with data in the format of Table 4.1 as the basis for a detector requirement. Fortunately, the calculation of D^* and the background photon flux density (BPFD, another parameter of interest to manufacturers) from the elements of Eq. 4.3 is simple:

Let $\bar{v}_c$ be the average frequency within the band $\Delta\bar{v}$ (suitably weighted if the band is more than 200–300 cm^{-1} wide). The average energy per photon is $E_p = hc\bar{v}_c$ J, where $hc = 1.9865 \times 10^{-23}$ J·cm. The total background at the detector $N_b = [N_2 + \Delta\bar{v}\cdot[n_3 + n_4 + n_5 + n_6)]/Q_d$ photons·s^{-1}. It can be shown (Kruse, 1977), assuming that the filter is at or near detector temperature and prevents longer wavelength radiation from reaching the detector, that the background-limited D^* is given by

$$D^* = Q_d/\{E_p[(2PN_b)^{1/2}]\} \text{ jones}$$

and remembering that $P = 1$ for a PV and 2 for a PC detector.

It also follows that $BPFD = N_b/A$ photons$\cdot$s$^{-1}\cdot$cm^{-2}, where A is the detector area.

4.1.7. The Impact of Pointing Jitter

If image motion (from any origin—jitter, smear, atmospheric turbulence, etc.) is significant, it induces an added amplitude modulation to the interferogram that will appear as noise in the resultant spectrum. The modulation is proportional to the total signal reaching the detector. If the proportionality factor is s', where $0 \leqslant s' \leqslant 1$, then for the FTS a term

$$s'(n_5 + n_6)\Delta\bar{v}\cdot t \tag{4.5}$$

must be added to the denominator. For a dispersive system, the term is

$$s'(n_5 + n_6)\delta\bar{v}\cdot(t/m) \tag{4.6}$$

where, in either case,

$$s' \propto \frac{1}{t}\int_0^t \frac{1}{f_2 - f_1}\int_{f_1}^{f_2} S(f)\exp[-2\pi ift]\, dt\, df \tag{4.7}$$

Here $S(f)$ is the *amplitude* spectrum of the scene modulation, assumed bounded by temporal frequencies f_1 and f_2, and t is the total integration time. However, with an FTS it is possible to employ a stratagem (to be explained in more detail later): if the amplitude spectrum is well known (and thus f_1 and f_2), the sampling interval and scan rate can be adjusted so that the modulation frequencies are largely (or entirely) shifted into a region of zero signal. This must be done with great care but is commonly used by astronomers to put noise due to atmospheric turbulence outside the desired spectral range and is also frequently used to evade problems of electrical interference from AC lines and switching power supplies.

Under these conditions, the composite expression for the SNR of an FTS becomes

$$\text{SNR} = \frac{\delta\bar{v}\cdot\Phi(n_5 + n_6)t}{[(N_1 + N_2)t + \Delta\bar{v}\cdot(n_3 + n_4 + n_5 + n_6)t]^{1/2} + s'\Delta\bar{v}\cdot(n_5 + n_6)t} \tag{4.8}$$

where s' is now to be interpreted as that portion of the signal modulation that falls within the occupied spectral interval. Of course, for uniform scenes (limb viewing, oceans, snow fields, etc.), $S(f)$ is zero for all f. The worst case is a "checkerboard" scene whose boundaries exactly match the detector elements. Analysis of *Landsat* Thematic mapper data suggests that, for typical scenes, $S(f) \propto 1/f$.

For the dispersive system, we get

$$\text{SNR} = \frac{\delta\bar{v}\cdot(n_5 + n_6)\cdot t/m}{\{[(N_1 + N_2) + \Delta\bar{v}\cdot n_3 + \delta\bar{v}\cdot(n_4 + n_5 + n_6)]t/m\}^{1/2} + s'\,\delta\bar{v}\cdot(n_5 + n_6)(t/m)}$$

(4.9)

4.1.8. Discussion

It is quite evident that if $s'\,\Delta\bar{v}/\delta\bar{v}$ approaches 1, we will arrive at a condition where the SNR for the FTS *becomes independent of the signal*, a condition to be avoided at all costs because the "spectrum" generated will be pure garbage. It is for this reason that an FTS must be used in a "staring" mode with precision pointing. The dispersive system is much less susceptible to jitter-noise corruption, so can be relatively poorly pointed and used in push-broom and whisk-broom systems.

4.2. INTERFEROGRAM SAMPLING

A critical element in the operation of an FTS is proper control of the interferogram sampling. First and foremost, an FTS is not amenable to the use of an external clock for sampling because fairly massive optical elements must be moved and it is difficult to control the rate of change of optical path difference (RCOPD) to better than a few tenths of a percent, which is not nearly good enough. The solution is to allow the FTS to generate its own clock.

The means universally employed is to illuminate the optical system with light from a frequency-stabilized laser and to sense the output with detectors separate from the main infrared detectors. For a truly monochromatic input and a constant RCOPD, the output is a cosine wave of constant frequency, as we saw in Chapter 1. Most frequently used is a red He/Ne leser, for which $\bar{v}_c = 15798.0024\ \text{cm}^{-1}$, although ongoing developments in solid-state Nd:YAG lasers ($\bar{v}_c \approx 9397\ \text{cm}^{-1}$) may make them an attractive alternative in the future. While there are several ways of exploiting the cosinusoidal output, they are all essentially equivalent to sensing the zero crossings of the cosine wave.

4.2.1. Sampling Theory

The logic of interferogram sampling is based on an extension of the Shannon sampling theorem (Shannon, 1949; Kohlenberg, 1953). The extended theorem shows that a spectrum that is nonzero only in the interval $\bar{v}_2 \geqslant v \geqslant \bar{v}_1$ cm^{-1} can be *exactly* (except for noise) reconstructed from an interferogram sampled at intervals of path difference δx cm given by

$$\delta x \leqslant 1/[2(\bar{v}_2 - \bar{v}_1)]; \qquad \bar{v}_2 > \bar{v}_1 \geqslant 0 \qquad (4.10)$$

provided that $\bar{v}_2/(\bar{v}_2 - \bar{v}_1)$ *is integral.* This integral ratio is commonly called the *alias order* (by analogy to the orders of a diffraction grating).

Note the insistance on *exact*. Sampling, when done correctly, does not result in an approximation: the reconstruction is perfect. A simple way of looking at it is to imagine that the function is periodic over the interval $\bar{v}_2 - \bar{v}_1$. We saw in Chapter 1 that such a function can be represented by a Fourier series (rather than a Fourier transform). A series is inherently a set of discrete values (i.e., samples).

A minor point to be noted is that the periodicity forced on the data by this type of sampling is actually over a period of $2(\bar{v}_2 - \bar{v}_1)$ so that, after transformation, data falling into regions of even alias order appear frequency reversed. This is trivially corrected in the software, but accounts for the alternations which will appear in subsequent illustrations.

The sampling requirements can be reversed: given a sampling interval δx, a set of contiguous spectral intervals is defined as shown in the accompanying tabulation.

Alias Order	Frequency Interval
1	$0 \to \bar{v}_0$
2	$\bar{v}_0^- \to 2\bar{v}_0$
3	$2\bar{v}_0 \to 3\bar{v}_0$
$\vdots$	$\vdots$
J	$(J-1)\bar{v}_0 \to J\bar{v}_0;$ $J = 1, 2, 3, \ldots$

Given such an interval $(J-1)\bar{v}_0$ to $J\bar{v}_0$, there follows the secondary requirement that the system filtering (optical and electronic) prevent optical frequencies less than $(J-1)\bar{v}_0$ cm^{-1} or greater than $J\bar{v}_0$ cm^{-1} from traversing the FTS system. If this is not accomplished, the phenomenon of *aliasing* occurs. Aliasing

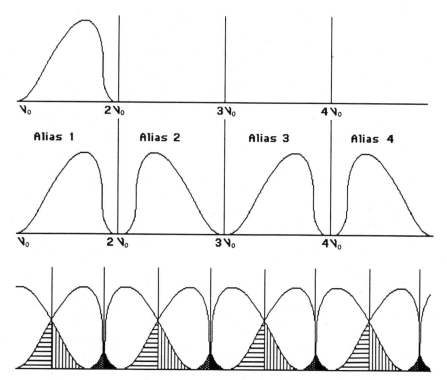

Figure 4.5. Interferogram sampling: *top*, input spectrum; *middle*, output spectrum with proper sampling; *bottom*, effect of improper sampling (aliasing).

manifests itself by reflecting those parts of the spectrum that fall outside the boundaries back into the spectrum. At the very least, the effect is to distort the data; at worst, the results are useless. The effect is illustrated in Fig. 4.5.

Now, as discussed above, the sampling interval is defined by the zero crossings of the cosine wave generated by a stabilized laser. Note that there is no intrinsic requirement that the sampling occur at *every* zero crossing. Depending on the filter characteristics, it may be possible to sample every Kth crossing ($K = 1, 2, 3, \ldots$). The zero crossings (in any given sense) occur precisely at intervals of one wavelength of the laser. If the laser frequency is $\bar{\nu}_c$, the allowable frequency intervals are therefore

$$\bar{\nu}_{\min} = (J - 1)\bar{\nu}_c/(2K)\,\mathrm{cm}^{-1}$$

Table 4.2. Allowed Spectral Intervals for a He/Ne Laser-Controlled FTS[a]

Sampling Interval (K)
(He/Ne Laser Wavelengths)

Alias Order (J)	1	2	3	4	5	6	7	8	9	10	11	12	13	14	15	16	17	18	19	20
0	0.0	0.0	0.0	0.0	0.0	0.0	0.0	0.0	0.0	0.0	0.0	0.0	0.0	0.0	0.0	0.0	0.0	0.0	0.0	0.0
1	7899.0	3949.5	2633.0	1974.8	1579.8	1316.5	1128.4	987.4	877.7	789.9	718.1	658.2	607.6	564.2	526.6	493.7	464.6	438.8	415.7	395.0
2		7899.0	5266.0	3949.5	3159.6	2633.0	2256.9	1974.8	1755.3	1579.8	1436.2	1316.5	1215.2	1128.4	1053.2	987.4	929.3	877.7	831.5	789.9
3			7899.0	5924.2	4739.4	3949.5	3385.3	2962.1	2633.0	2369.7	2154.3	1974.7	1822.8	1692.6	1579.8	1481.1	1393.9	1316.5	1247.2	1184.8
4				7899.0	6319.2	5266.0	4513.7	3949.5	3510.7	3159.6	2872.4	2633.0	2430.5	2256.9	2106.4	1974.8	1858.6	1755.3	1662.9	1579.8
5					7899.0	6582.5	5642.1	4936.9	4388.3	3949.5	3590.5	3291.2	3038.1	2821.1	2633.0	2468.4	2323.2	2194.2	2078.7	1974.7
6						7899.0	6770.6	5924.2	5266.0	4739.4	4308.5	3949.5	3645.7	3385.3	3159.6	2962.1	2787.9	2633.0	2494.4	2369.7
7							7899.0	6911.6	6143.7	5529.3	5026.6	4607.7	4253.3	3949.5	3686.2	3455.8	3252.5	3071.8	2910.2	2764.6
8								7899.0	7021.3	6319.2	5744.7	5266.0	4860.9	4513.7	4212.8	3949.5	3717.2	3510.7	3325.9	3159.6
9									7899.0	7109.1	6462.8	5924.2	5468.5	5077.9	4739.4	4443.2	4181.8	3949.5	3741.6	3554.5
10										7899.0	7180.9	6582.5	6076.2	5642.1	5266.0	4936.9	4646.5	4388.3	4157.4	3949.5
11											7899.0	7240.7	6683.8	6206.4	5792.6	5430.6	5111.1	4827.2	4573.1	4344.4
12												7899.0	7291.4	6770.6	6319.2	5924.2	5575.8	5266.0	4988.8	4739.4
13													7899.0	7334.8	6845.8	6417.9	6040.4	5704.8	5404.6	5134.3
14														7899.0	7372.4	6911.6	6505.1	6143.7	5820.3	5529.3
15															7899.0	7405.3	6969.7	6582.5	6236.1	5924.2
16																7899.0	7434.4	7021.3	6651.8	6319.2
17																	7899.0	7460.2	7067.5	6714.1
18																		7899.0	7483.3	7109.1
19																			7899.0	7504.1
20																				7899.0

[a]Intervals go from $(J - 1)$ to J and are labeled by J (e.g., "alias order 1" covers the interval between row 0 and row 1).

and

$$\bar{\nu}_{max} = J\bar{\nu}_c/(2K)\,\text{cm}^{-1}$$

where $J = 1, 2, 3, \ldots$ (the alias order) and $K = 1, 2, 3, \ldots$ (the sampling index).

The benefit of this type of undersampling is that, provided the rules are obeyed, it offers a form of lossless data compression. That is, the data rate and volume are both decreased without any loss of information.

Table 4.2 shows the resultant allowed spectral intervals as a function of J and K, assuming a He/Ne laser. For signal chain design purposes, these entries can be converted to electrical frequencies through the transformation equation (Eq. 1.8): $f = 2\bar{\nu}V\,\text{Hz}$. For example, suppose that we wish to achieve a spectral resolution of $0.083\,\text{cm}^{-1}$ and that the permissible scan time is 1 s (both very plausible numbers for a downlooking atmospheric remote sensor). The required path difference is $8.447\,\text{cm}$ and the RCOPD $(2V)$ is obviously $8.447\,\text{cm}\cdot\text{s}^{-1}$.

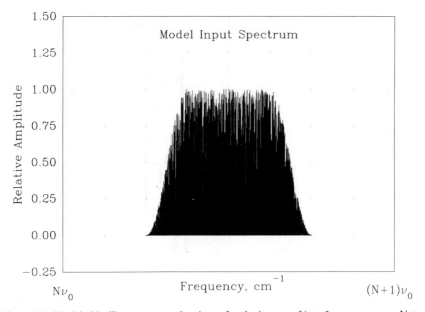

Figure 4.6. Model "ideal" spectrum used as input for the impact of interferogram errors. Note that this and subsequent spectral illustrations are shown as bar charts to emphasize the discrete nature of the data. However, the bars are so dense (8192 points are displayed across the frame) that the impression is one of solidity.

The clock frequency generated by the He/Ne laser is therefore

$$f_1 \approx 15798 \times 8.447 \approx 133.4 \, \text{kHz}$$

and from the data in Table 4.2 we see that the infrared signal chain (including the detector) may well need to encompass electrical frequencies as high as 67 kHz. This, too, can create design problems.

4.2.2. A Brief Digression

In the following sections, several examples of spectral distortion resulting from errors in the interferogram will be presented. In every case, the "ideal" spectrum is that shown in Fig. 4.6. The procedure was to set up the model spectrum of 8192 points (deliberately strongly structured to emulate real data), take its Fourier transform (i.e., generate a model interferogram, also of 8192 points; see Fig. 4.7) and retransform into spectral space after various types of interferogram error were simulated. Note that, as for any numerical model, the

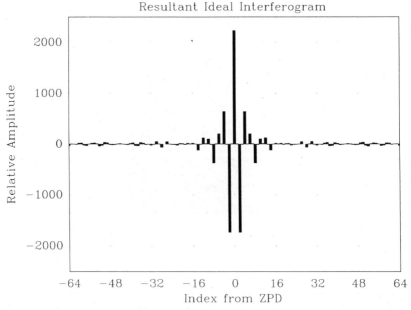

Figure 4.7. Central region of the theoretical interferogram obtained by a discrete Fourier transformation of Fig. 4.6.

results apply only to the specific circumstances assumed. Furthermore, for the purposes of illustration, the distortions applied were gross. Generalization therefore requires both caution and many such trial computations and analyses for the particular system.

4.2.3. Sampling Position Accuracy

The required accuracy of location of sampling points is severe although, strictly speaking, it is the *knowledge of the spacing* between points that is most critical. The absolute location is a secondary issue that can be dealt with in the data processing.

Two types of error must be avoided—periodic and random. Periodic errors generate "ghosts" within the spectral interval (Fig. 4.8). Detailed computation and simulation suggests that periodic variations in the sampling interval must be kept to less than 10 Å (1 nm) peak to peak. The tolerance for random error is a little greater: 20 Å (2 nm) RMS. *Bear in mind that both these numbers represent only a few atomic diameters!* It is in this requirement that the most

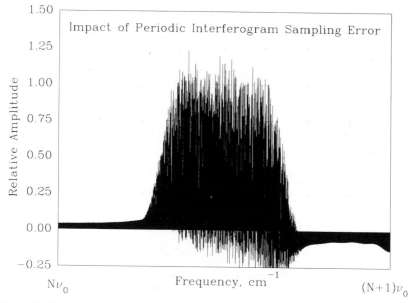

Figure 4.8. The result of 20% periodic sampling position errors in the interferogram of Fig. 4.7 and the impact on the spectrum (cf. the ideal case—Fig. 4.6).

serious difficulties of Fourier transform spectrometry lie and also indicates why sampling by an external clock is inadequate (Zachor et al., 1981).

4.2.4. Sampling Strategy

Optical filters do not have sharp cutoffs. That is, a filter of nominal bandwidth Δv will have measurable transmittance well beyond this limit (Fig. 4.9). Indeed, it becomes a matter of judgment to decide at what points the transmittance becomes negligible. Values of 200–300 cm^{-1} beyond each half-power point are not uncommon. Thus the allowed spectral band must be at least as wide as the band-pass plus this added range. It is also very useful for diagnostic purposes to have a "zero signal" region on either side of *that* extended range (a few tens of cm^{-1}), although this is not mandatory. If, however, intensity modulation is possible, then, as suggested earlier, one can modify the sampling strategy to alleviate the problem. This is best illustrated with a numerical example.

Assume that we are using the filter of Fig. 4.9. Its 50% (of peak) transmittance points are about 2370 and 3040 cm^{-1}. However, the transmittance remains

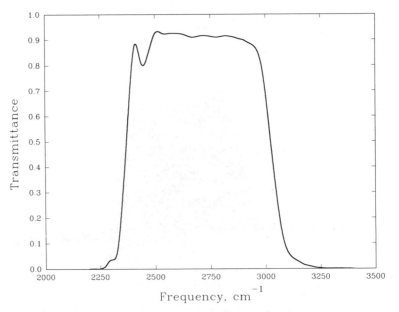

Figure 4.9. Transmittance curve for a typical band-pass filter used in an FTS.

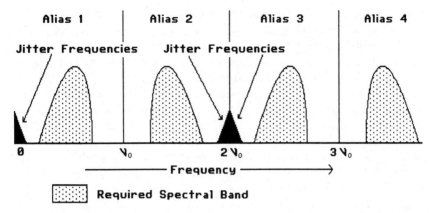

Figure 4.10. Strategy for reducing the effects of pointing jitter on an FTS.

above 0.1% between 2250 and 3280 cm^{-1}. Examination of Table 4.2 suggests that, for safety, one should therefore select $K = 4$ and $J = 2$, giving an allowable band of 1974.8–3949.5 cm^{-1}. If the scan speed is the same as our previous example (8.447 cm·s^{-1}), the electrical bandwidth of the signal leaving the detector will fall between 19 and 28 kHz. With an allowable band (in these units) of 16.7 to 33.4 kHz, we have a "pad" of at least 2 kHz at the low-frequency end.

Now, the nature of the sampling is such that the alias chosen ($J = 2$) produces an output *identical* to that which would appear if we were to translate the band-pass down 1974.8 cm^{-1} ($\equiv 16.7$ kHz) so that the lowest frequency becomes zero (in both systems of units). The "pad" now ranges from 0 to 2 kHz. There-fore, any signal modulation that falls within this range is "aliased" out of the wanted spectral band and reduces the problem. It is generally true that most mechanical disturbances and atmospheric turbulence effects would usually cover a much smaller range (0–300 Hz would be typical), so the strategy might be expected to work (see Fig. 4.10). Note, however, that it is real energy that is being extracted from the wanted spectrum. The consequence is that the radiometric calibration will be compromised if the "extracted" power exceeds a few tenths of a percent. Thus avoidance of signal modulation effects is still to be preferred. This topic will be revisited later.

4.2.5. Sampling Methodology

There are two fundamentally different approaches to the control of an FTS. The first, and simplest, is the technique that has been assumed throughout this book: a mirror is caused to translate at constant velocity, resulting in a continuous output at the detector, which is then appropriately sampled. This is usually called *rapid*, or *continuous, scanning*. The second approach steps the mirror between sample points as quickly as possible and then stops, holding the mirror position for the desired integration time. This method, called a *step-and-lock system*, has several advantages. First and foremost, the sampling positional accuracy is 10–100 times better than for rapid scanning; RMS accuracies less than 1 Å are routinely reported. Second, because the system is static during the integration, signal modulation is performed by oscillating the path difference through a small range (actually $\pm \lambda_m/4$, where λ_m is the mean wavelength of the infrared band). Since this frequency is now decoupled from the scan speed, the user is free to make the modulation frequency any convenient value (a few hundred hertz to 1 kHz is typical). This immediately gets around the frequency response problems of the rapid scanner and further-more totally eliminates electronic phase dispersion problems (more on this later). Third, the path difference modulation effectively differentiates the inter-ferogram so that ZPD, instead of being a maximum or a minimum, is now a zero crossing (which is always easier to find than a peak). Furthermore, such differentiation goes a long way toward eliminating problems of baseline drift of the interferogram (although few modern systems of any type suffer from this problem). Incidentally, the transformation of such an interferogram can be done directly: one merely takes a sine (rather than a cosine) transform. Fourth, the approach lends itself to *absolute* position control. That is, one can offset the control so that there is a sample point precisely at ZPD. While (as we have seen), the lack of absolute position control can readily be overcome in the analysis software, having absolute control lends itself well to real-time data analysis. Particularly for astronomical observations, this is a very useful feature. The final point is more subtle, but potentially the most important. As we have seen, an FTS can be configured as a spectral camera using two-dimensional (2-D) detector arrays. What was not then discussed was how to get the data out. Since a rapid scanner (a) inherently requires a continuous output and (b) this output must be sampled by the internal (laser-generated) clock, it follows that a rapid scanner *cannot use a multiplexed focal plane*, quite apart from the frequency response problems. Consequently, each element of the 2-D array must have its own signal chain. With a 32 × 32 array (by no means large even by present standards), one is faced with 1024 parallel signal chains, a not insignificant problem in terms of power dissipation, cooling, and data handling. A step-and-lock system is not so limited and, indeed, could be used in a DC mode if the detector properties permitted. The step-and-lock system is perfectly compatible with multiplexed detector arrays and is therefore better suited for the imaging mode.

If step-and-lock systems are so much better than rapid scanners, why are there so few of them (and most of those in France)? First, the servo design is nontrivial and demands engineering skills not usually possessed by spectrometer builders. Second, and most important, the need to accelerate and decelerate fairly massive optical elements means that it is very difficult to achieve sample rates much above 100–200 per second with a reasonable duty cycle (> 50%, say). That is, for a step-and-lock system, the time spent moving between lock points is essentially "dead" time. As we have already seen, however, an orbital remote sensor demands rates at least 10–100 times greater. Consequently, the only step-and-lock systems currently in use are in the laboratory and at astronomical telescopes, where sheer speed is not a driving constraint.

In any event, we are now approaching the point where we can begin to diagram a conceptual system. Figure 4.11 shows the elements of a generic FTS remote sensor that warrant discussion in some detail. Note, however, that this

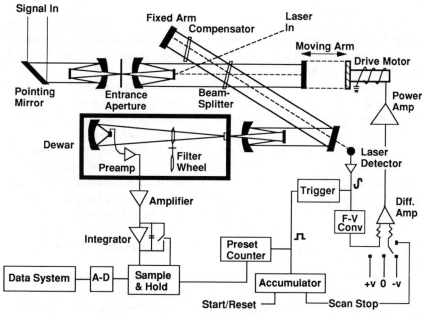

Figure 4.11. A generic FTS remote sensing system.

is not to be construed as a *design*. Indeed, as diagrammed, this system probably would not work!

At the top left is a pointing mirror. Since a stable input must be maintained for at least the time required to make one interferogram, the mirror should preferably be provided with an auxiliary reference. This might be an imaging system to provide direct target feedback or, for observations of the limb, a horizon sensor. Alternatively, a more general approach would be to use a star tracker and possibly an inertial reference unit (three-axis gyro). Following the pointing mirror is a telescope (collector). The system sketched has unit magnification for which, were the system in near-Earth orbit, the ground sampling distance (GSD) would lie in the range of hundreds of meters to a few kilometers. If the requirements demand a smaller GSD, a larger telescope might be used. However, if the telescope becomes much larger than about 30 cm aperture, the external pointing mirror becomes unreasonably large and one might consider pointing the entire system (thus eliminating the pointing mirror).

At the focus of the collector is an entrance aperture (field stop) whose primary function is stray light control. Limb sounders, for example, face very strong radiance gradients in the atmosphere, so a high degree of off-axis rejection is essential.

Behind the aperture is a collimator, shown as a Cassegrain system. Since this puts a "hole" into the middle of the beam, it also provides a convenient point for injecting the light from the control laser. Many other configurations are, of course, possible.

The collimated beam travels to the beamsplitter, now at last shown more realistically as having a finite thickness. Assuming that the coating is on the front face, the transmitted beam would make three passes through the material while the reflected beam (shown going to the fixed arm) makes only one en route to the detector. It is therefore standard practice to put an identical uncoated element called a *compensator* into the reflected beam to ensure equality. It transpires that the alignment (tilt) of the compensator can be quite critical. Another point to note is that we have gone away from the original Michelson interferometer configuration—the arms are no longer perpendicular. The shallower the angle, the smaller the beamsplitter and compensator. Since these are usually the most expensive optical elements, anything that reduces their size is beneficial.

In this schematic, we show only one moving arm. It is, however, quite common to move both (in opposition, of course), providing double the optical path difference (OPD) for a given physical travel. The drive motor is shown as a simple linear electromagnetic motor. While these are perhaps the most widely used drivers, systems have often been built using precision lead screws (as in a lathe) and even hydraulic motors. The moving arm(s) must be mounted in some device to constrain motion to one linear dimension. Two-rail systems using recirculating ball bushings or gas or oil bearings are quite common for large movements. For low-resolution systems requiring motions of a centimeter or so, flexure systems are very popular, being free of friction and wear (other than fatigue). For *very small* motions (millimeters), the motor and mount are often combined in a modified voice coil.

The outgoing beam is intercepted by a bending flat (mainly to save space in the figure!) with a hole to permit convenient extraction of the control laser beam. The electrical output (our sinusoid) is amplified and shown split two ways. One direction goes to a circuit that provides an output voltage proportional to the laser frequency. When compared to a constant voltage, the output provides the control signal for a velocity feedback to the drive motor. The system is stopped simply by setting the comparison voltage to zero, and the scan is reversed by reversing the comparison voltage. As sketched, this system has no directional sensitivity because the output of the frequency-voltage converter is independent of direction. There are, however, several ways of providing such knowledge either directly or through auxiliary subsystems.

The other direction for the laser output is to a trigger circuit that generates a pulse whenever the sine wave crosses zero (usually in one sense only to avoid problems of hysteresis in the circuitry). This pulse train is also split. One part goes to an accumulator that is set to zero at the beginning of a scan. When enough pulses have been counted (i.e., OPD is sufficient), a "stop" command is issued to the drive circuitry (and probably also to the data system). The other part goes to a programmable present counter that operates as a "divide-by-K" circuit, where K is the sampling interval chosen (via Table 4.2) to match the required spectral band. That is, one pulse is emitted for every K pulses in.

Meanwhile, back at the bending flat, the infrared beam has gone to a precondenser. Its function is to reduce the beam size so that the entrance window to the detector Dewar system can be as small as possible. This is done both to reduce heat loads on the Dewar system and for safety: there may often be a 1-atm pressure differential across the window. Not shown, but often used, is a field lens to image the beamsplitter onto the filter (reduces filter size). While it is tempting to make the field lens double as the window, the temptation should be resisted. Not only are distortions likely but also the internal elements of a Dewar system shrink quite alarmingly on cooling. The field lens and all subsequent elements should therefore be firmly coupled to the cold head and not the near-ambient Dewar wall.

Immediately after the Dewar entrance is shown another field stop. For best performance, this stop should (a) be the limiting system stop (i.e., the smallest in terms of field of view) and (b) be at or very near to the detector temperature. Similarly, the band-pass filters (shown on an internal wheel) must also be at a low temperature. Otherwise, they become noise sources rather than noise limiters.

A final condenser puts the infrared radiation onto the detector element. Infrared detectors are usually quite small [typically 0.1–1.0 mm square or diameter], so they are often mounted in the incoming beam as shown. Off-axis elliptical condensers are also quite common.

The detector output goes to a preamplifier mounted within the Dewar, both to reduce the likelihood of corruption of the very small voltages typical of detector outputs and for better noise performance. Operating electronics at low temperatures is not trivial. Components must be specially selected and mounted and only certain types of transistor (usually field effect transistors) can be used. The amplified (or impedance-matched) signal can then be delivered to more conventional electronics which might, for a rapid-scanning FTS, include an electronic band-pass filter to match the optical filter.

The system shown features a gated integrator that is cleared by the end of the previous sampling and read, with a simple-and-hold circuit, by the next sample pulse. The sample-and-hold circuit, in turn, feeds an analog-to-digital converter (A–D). Of course, many other types of integrator exist that are far superior to the one illustrated. A particularly elegant approach that also avoids the use of an A–D converter is to feed the analog signal to a voltage-frequency converter followed by an up-down counter. The value on the counter at the end of intergration is the digitized output.

Beyond this point, we enter the digital domain, where almost any standard data recording system can be used. Not shown in this outline is the overall system controller that sets the system parameters and activates the initial start, because this can range from a purely manual operation to a totally pre-programmed and automatic system.

4.2.6. Symmetric and Asymmetric Scanning

In order to recover a spectrum, the interferogram must have been sampled from zero path difference (ZPD) out to some maximum path difference (MPD). It is therefore conventional to design the scan system so that a few hundred samples are acquired on the other side of ZPD. It can be shown that, provided the sample spacing is correct, a new interferogram can be constructed from the measured data that places a sample *exactly* at ZPD. The Fourier trans-formation process can then proceed. However, the points on the "wrong" side of ZPD are effectively wasted (a loss of duty cycle). Many systems therefore choose to scan from $-$MPD to $+$MPD so that ZPD occurs near the middle of the scan. The resampling can still be accomplished, and no sample points are lost. That is, symmetric scanning is slightly more efficient than the asymmetric approach. The cost is, of course, more mirror travel and a physically larger system, so the designer must make a trade-off between efficiency, on the one hand, and size and power consumption, on the other. Both methods are in common use.

4.3. OPTICAL CONFIGURATIONS

In this section, we take a closer look at the optical configuration of the FTS. To this point, we have considered only the Michelson interferometer with only modest design changes from that used by A.A. Michelson himself in the nineteenth century. As we shall see, this is neither the only nor necessarily the best option.

4.3.1. The Michelson FTS

A optical configuration of a classic Michelson-type FTS might look rather like Fig. 4.11. The great advantage of this approach is that it uses the absolute minimum number of surfaces. Its great defect is the need to maintain the parallelism of the moving mirror(s). The requirement is that departures from parallelism be less than $\lambda/8$, where λ is the shortest wavelength of interest. This is very difficult to achieve and becomes worse the greater the movement. Since a typical modern remote sensing FTS will require a maximum OPD of tens of centimeters, the configuration is now rarely employed. When it is, it will usually have added to it a relatively complex active mirror alignment system (usually based on illuminating the mirror periphery with two or more auxilliary laser beams). Since this alignment system constitutes a single-point failure mode, it can hardly be recommended for use in remote environments.

4.3.2. Retroreflectors

The most important modification that can be made to the simple system is to replace the plane mirrors by *retroreflectors*. A retroreflector is, as its name suggests, an optical system that returns light on a path parallel to itself largely independent of its own orientation. Furthermore, it offers the possibility of separating the incoming and outgoing beams, permitting detection of that part of the signal that, in a Michelson FTS, returns to the source. Since this now requires two detectors whose noise adds randomly, the gain in SNR is at most $2^{1/2}$ but is nonetheless worthwhile. Yet another benefit will be described in the next section.

The simplest type is the *cat's-eye retroreflector* (Fig. 4.12a). It is constructed from a parabolic primary with a small secondary mirror at the focus. Parallel light entering the system is focused onto the secondary, returning to the primary and emerging once more parallel. Although even with a curved secondary the system has a limited range of angles over which it can operate, these are far larger than any potential "wobbles" in the drive and suspension system. It transpires that the only critical dimension is the primary–secondary spacing (Beer and Marjaniemi, 1966), so this is usually maintained through the use of a low-expansion spacer such as a fused silica or Invar tube. An added benefit of the cat's-eye is that the secondary mirror is readily mounted on a piezoelectric element that can be used for high-frequency modulation of the path difference and/or as a "tweeter" on the path difference servo control. Cat's-eyes are widely used.

An alternative approach uses a *cube-corner reflector*, a 2-D rendition of which is shown in Fig. 4.12b. It is actually constructed from three mutually perpendicular plane mirrors in a monolithic assembly. Its great virtue is that it is considerably shorter and less massive than a cat's-eye (saving weight and power). Its defects are (a) it is extremely difficult to make a cube-corner reflector to the required precision (typically no more than 1 arc second from perfect orthogonality), so they are very expensive; (b) the three successive 90° reflections can introduce unwelcome polarization into the beam; (c) the internal edges, if not razor sharp, can create significant stray light problems since they are directly in the beam. Nevertheless, as cube-corner reflectors' availability improves, they are becoming increasingly popular.

We now turn to the means of employing such elements in an FTS. For pictorial simplicity, we shall assume the use of cat's-eyes (it being almost impossible to portray the 3-D reflection pattern of a cube-corner in 2-D).

(a)   (b)

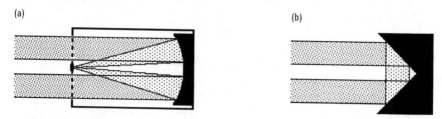

Figure 4.12. Sketches of (a) a cat's-eye retroreflector and (b) a cube-corner retroreflector (actually, a so-called roof mirror, since a true cube-corner cannot be drawn in 2-D).

4.3.3. The Connes-Type FTS

This configuration was developed by Pierre Connes (of CNRS, France) and his students between 1965 and 1975. In fact, he built his first prototype during a sabbatical year at JPL in 1966. The usual configuration is shown in Fig. 4.13.

The most obvious feature of the Connes approach is that not only are there two outputs, there are also two inputs. Let us first trace the left-hand input (labeled "1" in the figure).

The input beam strikes the beamsplitter, now physically separated from the recombiner. The two beams enter the retroreflectors and emerge displaced sideways, returning to the recombiner. Note that the plane of splitting is not the same as the plane of recombination. This is done to compensate for the shear introduced by the finite thickness of the splitter and recombiner (which must otherwise be identical). Shear between the beams is a problem because it reduces the area of overlap and hence the modulation efficiency. The required offset δ of the planes is a function of the thickness of the plates (d), their refractive index (μ), and the angle of incidence (θ). Specifically, the shear is zero when

$$\delta = d\{1 - \cos(\theta)/[\mu^2 - \sin^2(\theta)]^{1/2}\} \tag{4.11}$$

as is easily verified by trigonometry and Snell's law of refraction (see Fig. 4.14).

By conservation of energy, the sum of the two outputs must equal the input I_1 (in the absence of transmittance losses). If we identify the outputs as O_{1a}

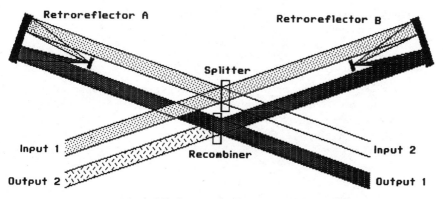

Figure 4.13. Layout of a Connes-type FTS.

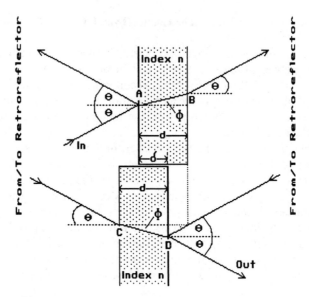

Figure 4.14. Eliminating beam shear in a Connes-type FTS.

and O_{2a}, then, following Eq. 1.13 for the monochromatic case,

$$O_{1a} = I_1[1 + \cos(2\pi\bar{v}x)] \tag{4.12a}$$

and, by definition,

$$O_{2a} = I_1[1 - \cos(2\pi\bar{v}x)] \tag{4.12b}$$

Detectors placed in each of the output beams and connected differentially provide an output

$$O_{1a} - O_{2a} = 2I_1\cos(2\pi\bar{v}x) \tag{4.13}$$

automatically removing the DC bias in the signal and doubling the output (but not the SNR because the noise of the two detectors adds incoherently). The gain is a factor of $2^{1/2}$.

The right-hand input I_2 follows a reverse path, so the sense of the outputs is reversed:

$$O_{1b} = I_2[1 - \cos(2\pi\bar{v}x)] \tag{4.14a}$$

$$O_{2b} = I_2[1 + \cos(2\pi\bar{v}x)] \tag{4.14b}$$

Obviously, if $I_1 = I_2$, the result is an unmodulated signal in both arms. This is both a curse and a blessing. It can be a blessing because if we are observing a source that "sits" on a background (as in the astronomical case), we can insert a nearby patch of sky into the other beam and thereby automatically subtract out the background (but *not* the background noise). This is very useful in astronomy because the background is always very much stronger than the source (sometimes by orders of magnitude) and can create problems of system dynamic range if it is not removed.

The curse is that the second input exists whether we like it or not. When observing extended sources like the Earth, one must ensure that the second input is properly controlled. One elegant solution for an orbital remote sensor, if it can be implemented, is to provide the second input with a view of cold space (3 K). This is so much colder than any conceivable source that one can put $I_2 = 0$ with negligible error. If that solution is not possible, a cold plate or cavity can be substituted, but its contribution may not be negligible (albeit calculable).

Another point about this configuration is that, although a retroreflector is insensitive to tilts, *lateral* shifts cause the distance between the input and output beams to change, thereby introducing another source of shear. The largest source of such shifts is not bearing play but nonparallelism of the optical axis and the mechanical ways on which the retroreflector moves. The coalignment therefore becomes very important.

A method exists for eliminating this problem, at the expense of introducing another (there's no free lunch!). We treat this method next.

4.3.4. The Compensated FTS

Figure 4.15 shows the concept of the compensated FTS (Beer, 1967). The emergent beam from the retroreflector is returned on its path by reflecting surfaces integral to the beamsplitter. Lateral shifts are totally compensated provided only that the angles that the reflecting surfaces make with the beamsplitter are identical.

The concept as drawn has two drawbacks. First, the beamsplitter is very thick and, furthermore, must be cemented. Cements absorb strongly in the infrared, so the approach is not useful much beyond $1 \mu m$. Second (and common to all compensated systems) is that, once more, half the signal returns to the source. A more practical scheme for infrared use is shown in Fig. 4.16 (Breckinridge and Schindler, 1981; note, however, that their discussion of the configuration of Fig. 4.15 is incorrect).

The surface returning the beams through the system is now separated from

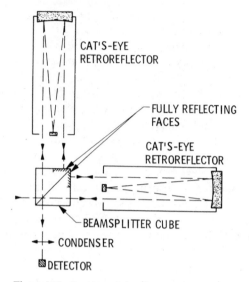

Figure 4.15. Concept of the tilt-compensated FTS.

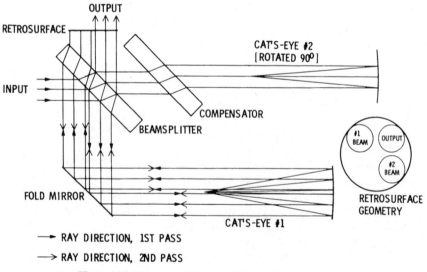

Figure 4.16. Practical realization of a tilt-compensated FTS.

the beamsplitter, reducing the thickness of transmitting material. Nevertheless, five passes are made through transmitting material and transmittance is certainly impaired. The retrosurface, as it is called, has the geometry shown on the right-hand side of the figure. The output beam is extracted through a hole. The system is proof against tilt of the retrosurface, because it affects both arms equally. Consequently, the configuration is often called *tilt compensated*.

4.3.5. Discussion

There are, of course, a remarkable number of other configurations that have been tried from time to time (Steel, 1971), but the three just discussed are the ones most widely used in remote sensing systems and we shall meet them again in Chapter 5. Each has its advantages and disadvantages, so the choice is a matter of judgment.

4.4. POTENTIAL PROBLEM AREAS

No technology is free of potential problem areas. We have already discussed some of these for the FTS, but there are more; none are insoluble. The important thing is to be able to recognize them as early in the design process as possible and proceed accordingly.

4.4.1. Phase Errors

To this point we have the assumption that an interferogram is a simple integration of cosine waves, looking somewhat like Fig. 4.17.

There are two phenomena that render this simple picture only an approximation. The first is optical dispersion. If the beamsplitter/recombiner and compensator are not identical (or the compensator is misaligned), the light in each arm of the FTS passes through a different optical thickness of material (*optical thickness* is the product of the physical thickness and the refractive index). Since refractive index varies with frequency, it follows that the path difference becomes frequency dependent. The consequence is most obviously seen at ZPD: instead of the elementary cosines being in perfect alignment (phase) at this point, a dispersive effect is seen that desymmetrizes the interferogram (Fig. 4.18). A similar effect can be induced by frequency-dependent phase changes on reflection at the beamsplitter/recombiner if the coatings have nonzero absorptance. The second source of phase dispersion is peculiar to the rapid-scanning FTS and is caused by frequency-dependent electrical

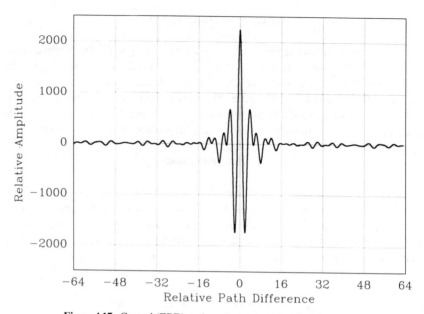

Figure 4.17. Central (ZPD) region of a typical ideal interferogram.

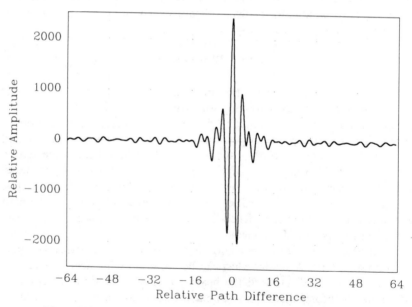

Figure 4.18. An interferogram displaying phase error (phase dispersion).

phase shifts in the detector signal chain. Such phase shifts are an inevitable consequence of finite electrical bandwidths. One of the added benefits of a step-and-lock FTS is that it is free of this second error.

Provided that the phase error is not too severe (less than 180°), an algorithm exists (Forman et al., 1966) that permits correction for this effect during data processing. The same algorithm also corrects for general shifts of the sample-point array and ensures that, before the actual Fourier transformation, a sample point lies exactly at ZPD. Nevertheless, it is clear that care must be taken to minimize the dispersive effect.

4.4.2. Channeling

A phenomenon unique to an FTS is channeling, which manifests itself as a near-sinusoidal modulation of the spectral continuum (typically a few percent peak to peak). While channeling is generated within the instrument and ought therefore to be removable by calibration (or some numerical tricks in data processing), it is universally found that (a) removal never works properly and (b) channeling changes with the character of the source under investigation; point sources generate much stronger channeling than do extended ones. Unfortunately, a remote sensor has no control over the character of the source, so the only recourse is avoidance. Examples of channeling are shown in Fig. 4.19.

The origin of channeling is optical interference between the back and front surfaces of transmissive elements (beamsplitter, windows, filters, etc.) in the optical path (or even between elements). It is therefore important to prevent this occurring by (a) "wedging" all such elements and (b) ensuring that no parasitic internal reflections reach the detectors. Since wedged elements also form dispersive prisms, care must be taken to ensure that such effects are compensating, not cumulative, through the optical train. Antireflection coatings can help, but not much because most FTS systems are required to operate over very wide frequency ranges. Coatings are usually effective only over very limited ranges.

It should also be noted that, while channeling is most severe if the parallel-sided elements are in a collimated beam, putting them in a convergent beam does not necessarily eliminate the effect. Detailed ray tracing is essential.

4.4.3. Vibration

We have already met one problem that can be generated by vibration: amplitude modulation of the input signal that can have a devastating effect on the

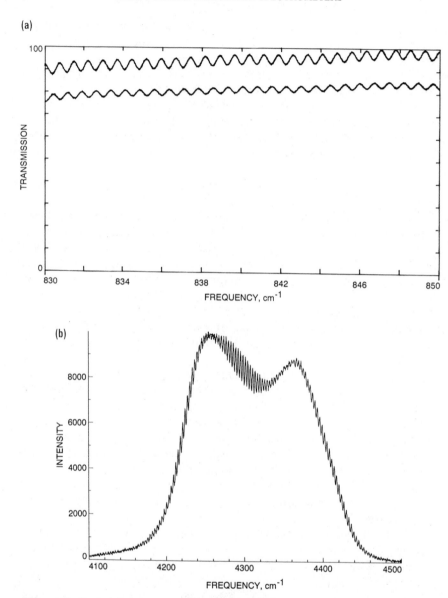

Figure 4.19. Examples of channeling in the spectral domain. (a) Non-reproducible channeling caused by the plane-parallel windows of an external absorption cell on successive days following removal and replacement. (Data courtesy of L.R. Brown.) (b) Channeling caused by the use of a plane-parallel filter within the detector Dewar. (Data courtesy of J.-P. Maillard.)

SNR and compromise the radiometric calibration. Equally bad is vibration coupling into the path difference control. Even though sampling is regulated by the control laser, vibration can introduce random sampling errors that, as we saw earlier, can also degrade the spectrum. The solutions are (a) very "tight" servo loops and (b) good vibration isolation from external sources. The latter can be a serious problem in unfavorable environments such as aircraft.

While not a solution to path difference modulation, some added protection against amplitude modulation can be achieved by *ratioing*. In general, one uses an auxiliary photometer or radiometer looking at the same source as the FTS but in a relatively broad optical bandpass. Dividing the FTS signal by this output can reduce the effects of amplitude modulation *provided that the SNR of the radiometer output is very much larger than that of the FTS*. If not, the exercise can be counterproductive. Note, however, that the electrical bandwidth of the radiometer need only encompass the expected modulation frequencies, offering a potential for longer instantaneous integration times and improved SNR.

This approach requires "stealing" some of the input energy. With a Connes-type FTS an alternative trick is often used. Referring back to Eqs. 4.12, 4.13, and 4.14, we see that if the outputs of the two arms are added, the sum equals the unmodulated input. Thus, dividing the difference by the sum of the same two detectors produces the same ratioing effect without any loss of signal.

4.4.4. Electromagnetic Interference

Electromagnetic interference (EMI) coupling into the signal chain at frequencies within the electrical passband is indistinguishable from the real signal. As we have seen, typical signal electrical frequencies for a remote sensor FTS lie in the kilohertz range. This is unfortunately also a typical frequency for a "switching" power supply. An FTS is *hypersensitive* to such effects and shielding well beyond that normally used on dispersive spectrometers is essential. The standard rules for avoiding "ground loops" must be scrupulously obeyed.

Some defense against EMI at frequencies outside the signal band (but one must watch out for harmonics!) can be had by employing the same strategy as is used for reducing amplitude modulation effects. A proper choice of sampling interval (i.e., a small degree of "oversampling") can be efficacious.

4.4.5. Signal Dynamic Range

The dynamic range D of an interferogram can be very large. To a fair approximation

$$D = S[(\Delta\bar{v}/\delta\bar{v})^{1/2}] \tag{4.15}$$

where S is the average SNR of the spectrum. Now, as we have seen, $\Delta\bar{v}/\delta\bar{v}$ can easily reach values of 10^4 or more so, for $S = 100$, we can easily find $D > 10^4$ and values of 10^7 are by no means unknown. This creates two difficulties: first, it is hard to find A–D converters with enough range (and those that have it are usually too slow); second, the requirements on nonlinearity are severe (Guelachvili, 1981).

The range problem is overcome by another stratagem. The critical issue is to avoid compromising the SNR through insufficient precision in the digitiza-

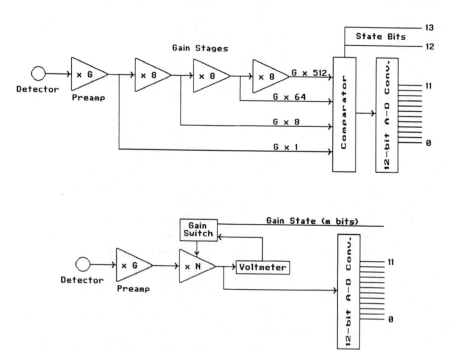

Figure 4.20. Some possible approaches to an FTS signal chain: *top*, cascaded system; *bottom*, autoranging system.

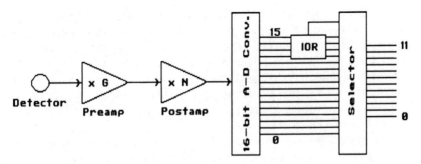

Figure 4.21. Alternative approach to Fig. 4.20 using a 16-bit A–D converter.

tion process. In order to get good characterization of the noise, at least 2–3 bits must be assigned to this task and enough over-range provided to avoid "overflow." A typical FTS therefore uses a 12-bit A–D converter that can generate data numbers (DN) in the range 0–4095. While this suffices for all but extremely high spectral SNRs (see below), it is not nearly enough for the

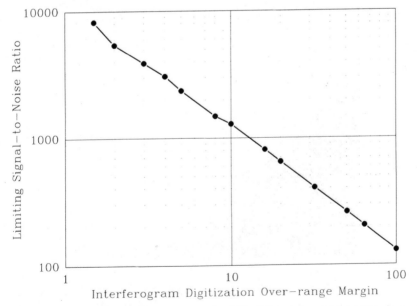

Figure 4.22. Limiting SNR using the approach of Fig. 4.21 on the model interferogram of Fig. 4.7.

interferogram. However, all that is required is either an automatic gain-changing system or a "cascaded" system as shown in Fig. 4.20. Of course, the gain conditions must be recorded, adding some overhead to the system (2 or 3 bits), but the consequence is improved speed (12-bit A–D converters are very fast) and linearity while effecting a form of lossless data compression, a very important issue that is considered later. Alternatively, if a good 16-bit A–D converter is available, an approach that selects only the 12 most active bits can be used (Fig. 4.21).

The limiting SNR available using this latter method was investigated using the model interferogram described earlier, adjusting the "system gain" to emulate various degrees of over-range capability (or margin). The results are shown in Fig. 4.22, where the SNR is taken to be

$$SNR = \frac{\text{peak of the input spectrum}}{\text{RMS difference between the output and input spectra}}$$

As can be seen, for this case an SNR in excess of 1000:1 is possible even for

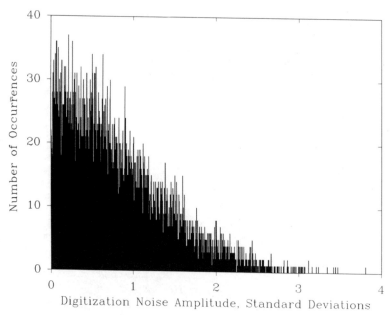

Figure 4.23. Amplitude distribution of the digitization noise engendered by the approach of Fig. 4.2.1.

an over-range margin as great as 10. Figure 4.23 shows the amplitude distribution from one of these trials; the shape is nearly gaussian, suggesting that with care it should be possible to co-add digitized interferograms or spectra and still obtain an improved SNR.

4.4.6. Signal Chain Linearity

If the relationship between "signal in" and "data numbers out" is nonlinear (or, more specifically, uncertain since it is in principle calibratable), the resultant spectrum has an indeterminate baseline (Fig. 4.24). This impacts both the radiometric calibration and the subsequent spectral analysis because the apparent spectral amplitudes are incorrect. Residual nonlinearity is therefore to be avoided by design; a typical requirement would be to have knowledge of the end-to-end system gain calibration to 0.5% or better. There are, however, conditions where this is difficult. For extremely strong signals such as are found in solar occultation sensors (see Chapter 3), the signal loading can be large enough to saturate the detector itself. This is rather difficult to calibrate

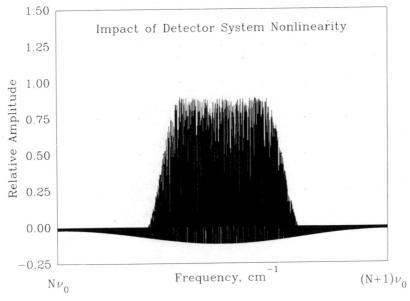

Figure 4.24. The impact of system nonlinearity (50%) on the model interferogram of Fig. 4.7 (cf. Fig. 4.6).

even in the laboratory because sufficiently bright sources are hard to come by (or even nonexistent).

4.4.7. Data Errors

An FTS is exceedingly unforgiving of data errors because *every* interferogram sample contributes to *every* spectral amplitude. Obviously, a 1-bit error in the final point at maximum path difference will have less impact than a similar error at ZPD; nevertheless, a general rule can be stated: the allowable bit error rate (BER) for an FTS is *zero* (Fig. 4.25). Now, real data systems (which, for a remote sensor, must include all transmission links) do not have a zero BER. The Tracking and Data Relay Satellite System (TDRSS), for example, has a BER of 10^{-5}. A typical interferogram of, say, 256 kilosamples digitized at 13–15 bits per sample comprises 10^6–10^7 bits, suggesting that, were nothing done, interferograms transmitted from space would have an average of 100 errors each. This is totally unacceptable. It follows, then, that sophisticated error-correcting codes are essential for a remote-sensing FTS. Read–Solomon encoding, for example, improves the TDRSS BER to 10^{-8}. Furthermore, experience

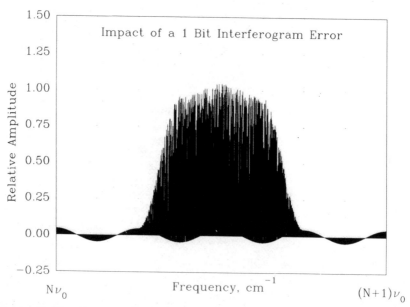

Figure 4.25. Impact of a 1-bit (sign) error at sample point 9 of the model interferogram of Fig. 4.7 (cf. Fig. 4.6).

shows that bit errors are usually nonrandom. They tend to come in bursts, so while any given interferogram may be worthless, reducing the losses to an acceptable value (a few percent, say) is quite feasible.

Recognizing that an interferogram has been "trashed" is often difficult. Gross errors (large spikes or total dropouts) can be recognized in the interferogram itself; more subtle (1-bit) errors can easily be overlooked. The final "proof of the pudding" would be the inability to extract believable results from the spectrum (which, in turn, implies that one knows what the "right" answer is). The real solution is to keep the BER as small as possible.

4.4.8. Data Rates and Volumes

The data rates and volumes for a remote sensing FTS can reach impressive values, stressing not only the transmission links but also the subsequent data

Table 4.3. Data Rates and Volumes for a Typical Remote Sensing FTS[a]

Spectral resolution	0.018	cm^{-1}
Max. path difference	33.8	cm
Max. path difference (He/Ne laser wavelengths)	533, 783	
Scan time	4	s
Scan speed (OPD)	8.447	$cm \cdot s^{-1}$
Digitization	13	bits/sample
Required spectral band	800–1050	cm^{-1}
From Table 4.2:		
Sampling interval (He/Ne laser wavelengths)	13	
Alias order	2	
Alias boundaries	607.6–1215.2	cm^{-1}
No. of interferogram samples[a]	41, 060	per detector
Free spectral range	607.6	cm^{-1}
Sampling rate (approx.)	10.26	kHz
Data rate	133.4	kbps/detector[b]
Data volume	534	kb

[a]*Notes:*
   No. of samples = max. path difference (PD)/sampling interval.
   Sampling rate = 2 × free spectral range × OPD scan speed.
   Data rate = sampling rate × digitization.
   Data volume = data rate × scan time.

   Volume doubles for symmetric scan (−max. PD to +max. PD).
   Volume and rate must be multiplied by the number of detector elements for an imaging FTS.
[b]kbs = kilobits per second.

processing. Indeed, with many current FTS systems it is not the spectrometer's ability to generate data that is the bottleneck but rather the inability of subsequent systems to cope, not excepting the investigator. While the speed of high-rate links and the availability of supercomputers is improving, at the end it is a human being who is required to make judgments about the results. A wide-open field for investigation is development of means for automating atmospheric spectral analysis. In this sense, the surface spectrometrists are better served—they already have many such tools available to them. Atmospheric spectrometry still has great room for improvement.

Table 4.3 shows the data rates and volumes for a typical orbital remote sensing FTS. Airborne and balloon systems would generally have rates an order of magnitude smaller because they are under less time pressure, but the volumes can be comparable.

CHAPTER

5

CASE STUDIES OF REMOTE SENSING FOURIER TRANSFORM SPECTROMETERS

In this chapter we shall give a brief overview of four existing and one planned FTSs. Each one is different, being intended for different purposes and operating in a different environment. The five are as follows:

- The *Voyager* IRIS FTS
- The Canada–France–Hawaii telescope (CFHT) FTS
- The *Spacelab 3* ATMOS FTS
- The Mark-IV Ballon/Aircraft FTS
- The *Eos* TES FTS

The rationale for choosing these five was that each one illustrates some particular points in FTS design and, just as important, enough information is available to permit useful discussion.

5.1. THE *VOYAGER* IRIS FTS

The *Voyager* IRIS (Infrared Interferometric Spectrometer) FTS had its inception in the mid-1960s. Early versions were flown on *Nimbus 3* and *4* for Earth observations, and its first planetary use (Mars) was on *Mariner 9* in 1971; the *Voyager* version to the outer solar system was launched in 1977. The Principal Investigator on all these experiments was Rudi Hanel of GSFC (Hanel et al., 1980).

5.1.1. Purpose

Voyager IRIS is a near-nadir sounder, measuring atmospheric composition, vertical temperature profiles, energy balance, and cloud properties of all the outer planets [Jupiter, Saturn (and one of its moons, Titan), Uranus, and

101

Table 5.1. Specifications for *Voyager* IRIS

Configuration	Michelson-type; continuous scan
Spectral coverage	180–2500 cm^{-1} (4–55 μm)
Max. resolution	4.3 cm^{-1} (apodized)
Max. OPD	2.3 mm
Scan time	46 s
Detector	Thermopile
Sensitivity (NESR at 500 cm^{-1})	5×10^{-9} W·cm^{-2}·sr^{-1}·(cm^{-1})$^{-1}$
Operating temperature	200 K
Beamsplitter	Cesium iodide
Beam diameter (internal)	3.3 cm
Telescope diameter	50 cm
IFOV (external)^a	4.4 mrad
System étendue	3.03×10^{-2} cm^2·sr
Path difference control	Auxiliary interferometer; neon lamp source
Weight	18.4 kg
Power	14 W
Data rate	1.12 kbps

^{a}IFOV = instantaneous field of view.

Neptune] except Pluto. IRIS has made a number of important discoveries during its 12-year career in space. Perhaps the most dramatic was the discovery of complex organic molecules in the atmosphere of Titan, which has given a major impetus for mounting the *Cassini* mission to Saturn with a Titan entry probe.

5.1.2. Specifications

The specifications for the IRIS FTS are shown in Table 5.1.

5.1.3. Optical Layout

Figure 5.1 shows the IRIS optical layout. Note that IRIS also incorporates a visible and near-infrared (0.33–2 μm) radiometer, not discussed here.

As can be seen in the figure, IRIS is a Michelson-type FTS. It is unusual in that it was decided to separate the visible control system from the infrared section through an auxiliary reference interferometer back to back with the main interferometer. Figure 5.2 shows the packaging of the system and Fig. 5.3 shows a closeup of the interferometer and its control system.

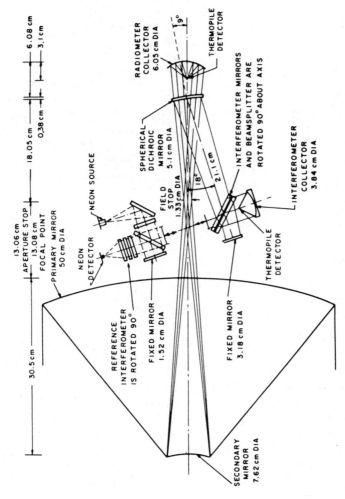

Figure 5.1. Optical layout of the *Voyager* IRIS FTS. (From Hanel et al., 1980. Copyright Optical Society of America; reprinted with permission.)

103

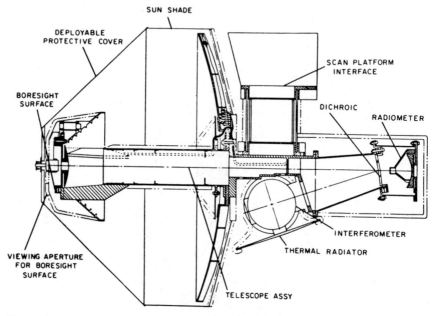

Figure 5.2. Packaging of the *Voyager* IRIS FTS. (From Hanel et al., 1980. Copyright Optical Society of America; reprinted with permission.)

5.1.4.   Discussion

It seems unlikely that IRIS will fly on any future missions. Indeed, an advanced version called MIRIS (Modified IRIS) was in development for *Voyager* at the time of launch but could not be readied in time. An even more advanced version called CIS (Composite Intrared Spectrometer), basically two IRIS interferometers back to back with extended coverage to 10 cm^{-1} and 10-cm path difference, is being considered for the *Cassini* mission to Saturn. Nevertheless, the original IRIS performed its assigned task for far longer than anyone expected and, for that alone, is worth consideration. A compendium of results is shown in Fig. 5.4.

5.2.   THE CANADA–FRANCE–HAWAII TELESCOPE FTS

On the 4.2-km-high summit of Mauna Kea, an extinct volcano on the island of Hawaii, is one of the largest collections of astronomical telescopes on Earth,

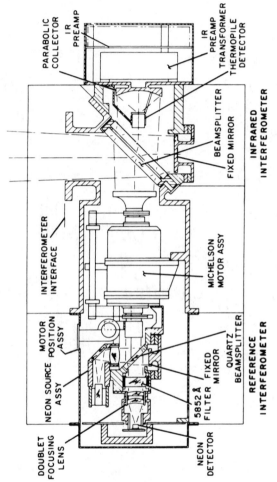

Figure 5.3. The IRIS interferometer section. (From Hanel et al., 1980. Copyright Optical Society of America; reprinted with permission.)

105

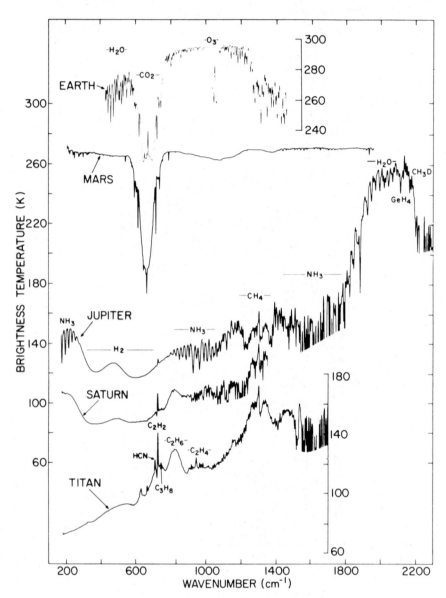

Figure 5.4. A set of spectra taken with various versions of the IRIS spectrometer. The spectrum of the Earth (over the mid-Atlantic Ocean) was acquired from *Nimbus 4* in 1970; that of Mars from *Mariner 9* in 1972; the spectra of Jupiter (1979) and Saturn and Titan (1980) are from *Voyager 1*. (From R. A. Hanel, 1981; reprinted with permission.)

with more to come before the end of the century. One of the smaller ones (really!) is a 3.6-m (142-inch) telescope belonging to a consortium of the governments of France and Canada and the University of Hawaii (usually known by its initials, CFHT). Mauna Kea is not only unpleasantly high, it is also one of the least humid places on Earth (which may astound anyone who has ever sweltered on a Hawaiian beach, but a meteorological phenomenon called the tropical inversion layer near 10,000 feet (~ 3000 m) ensures that all the moisture is trapped below this level), making it ideal for infrared astronomical observations if not for the observers. Among the instrument complement is an FTS used routinely to study the infrared spectra of stars and planets. The individual responsible for developing and operating this instrument is Jean-Pierre Maillard of the Institut d'Astrophysique in Paris (see Maillard and Michael, 1982).

The CFHT FTS is included in our discussion because, first, it uses some unique technology and, second, astronomy is surely the archetypical field of remote sensing. Furthermore, the environment of a high-altitude observatory is far from benign and makes some special demands on systems design.

5.2.1. Purpose

The CFHT FTS is designed to acquire near-infrared spectra of astronomical objects from the Cassegrain focus of the telescope. This means that it moves with the telescope and the gravity vector changes dramatically. Furthermore, the instrument is essentially inacessible in use, so all control is performed remotely from a room on a lower floor of the observatory. The reason for using the Cassegrain focus rather than the more usual coudé (fixed) focus is to improve efficiency (a factor of 1.5), to reduce the telescope background (a factor of 2.5), and to obtain a wider FOV (field of view). In addition, the CFHT Cassegrain focus housing contains a TV-based automatic guider that goes a long way toward eliminating pointing jitter, permitting the use of a smaller IFOV (instantaneous field of view) (down to 1.5 arc seconds on the sky) and thereby further reducing the background.

5.2.2. Specifications

The specifications for the CFHT FTS are shown in Table 5.2.

5.2.3. Optical Layout

The optical layout is shown in Fig. 5.5. The figure is inaccurate in that the beamsplitter and recombiner are actually separated as discussed in Chapter 4.

Table 5.2. Specifications for the *CFHT* FTS

Configuration	Connes-type (cat's-eye); step-and-lock system
Spectral coverage	$1800–25000\,cm^{-1}$ $(0.4–5.5\,\mu m)$
Max. resolution	$0.01\,cm^{-1}$ (unapodized)
Max. OPD	$60\,cm$
Scan time	0.1–50 steps/s (infinitely adjustable)
Detector	InSb PV (2) at 65 K
Sensitivity (NESR at $2100\,cm^{-1}$; 1-hr integration)	$2.6 \times 10^{-10}\,W\cdot cm^{-2}\cdot sr^{-1}\cdot(cm^{-1})^{-1}$ (background limited)
Operating temperature	Ambient ($\sim 0\,°C$)
Beamsplitter	Fused silica (2); calcium fluoride (interchangeable)
Beam diameter (internal)	$2.2\,cm$
Telescope diameter	$360\,cm$ ($336\,cm$ used)
IFOV (external)	Interchangeable; 1.5, 2.5, 5, 8, 12 arc seconds
System étendue	$\geqslant 3.7 \times 10^{-6}\,cm^2\cdot sr$
Path difference control	Stabilized He/Ne laser
Weight	$700\,kg$
Power	$2000\,W$
Data rate	700 bps (peak)

The three available splitter/recombiner pairs are mounted on a single optically worked fused silica block and interchange is totally automatic. No realignment is required. The packaging is shown in Fig. 5.6. The outer envelope is a tank approximately 2 m long by 1 m diameter and can be evacuated during storage (the instrument is used only intermittently). During operations, the instrument operates at ambient pressure.

5.2.4. Discussion

The CFHT FTS is undoubtedly the "state of the art" in astronomical FTSs. Its control system, based on the step-and-lock principle, has a loop bandwidth of about 1 kHz, permitting useful sampling rates of up to 50 samples per second. Furthermore, the control laser beam employs a sophisticated phase-modulation scheme that permits sampling at intervals as short as 1/8 wavelength of the control laser. This feature is very useful in astronomical investigations because the intense atmospheric background (beyond 3 μm) demands that spectra be acquired over quite narrow regions ($100\,cm^{-1}$ or so) and optimizing the sampling becomes more important. The system also features a near-real-time Fourier transformer that permits spectra to be displayed during acquisition.

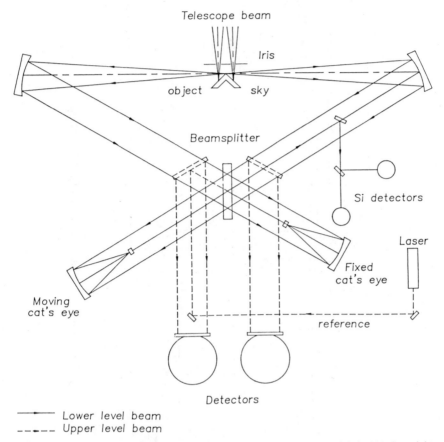

Figure 5.5. Optical layout (sketch) of the CFHT FTS. (From Maillard and Michel, 1982. Copyright Kluwer Academic Publishers/Reidel Publishing Co.; reprinted with permission.)

In addition, experiments are underway to assess its utility in a full 2-D imaging mode over a field of 24 arc seconds diameter, as discussed in Section 2.6. Examples of spectra acquired with this instrument are shown in Figs. 5.7 and 5.8.

5.3. THE *SPACELAB 3* ATMOS FTS

The shuttle-based mission *Spacelab 3* provided the first flight opportunity for the ATMOS (Atmospheric Trace Molecule Spectrometer) FTS and, incidentally,

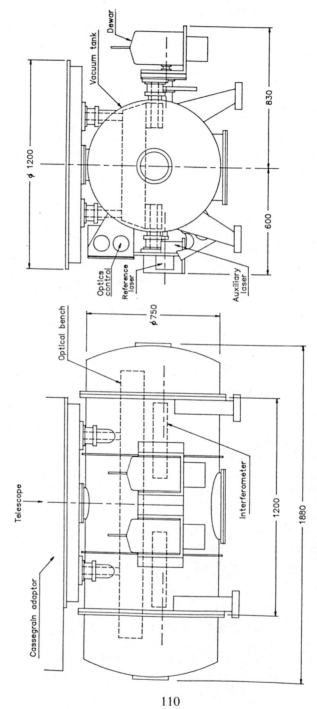

Figure 5.6. Packaging of the CFHT FTS. Dimensions are in millimeters. (Courtesy of J.-P. Maillard.)

110

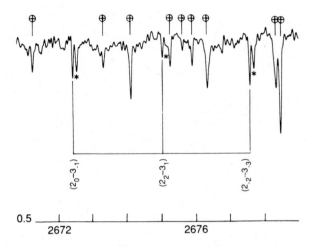

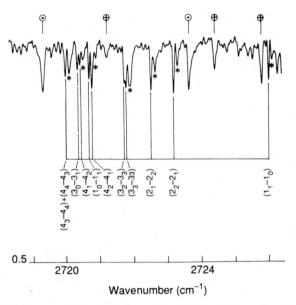

Wavenumber (cm⁻¹)

Figure 5.7. Portions of a spectrum of Mars ($\delta\bar{v} = 0.036\,\mathrm{cm}^{-1}$; Doppler shift, $0.125\,\mathrm{cm}^{-1}$) indicating the presence of HDO in the Martian atmosphere. The equivalent telluric line is indicated by an asterisk. (From Owen et al., 1988. Copyright American Association for the Advancement of Science; reprinted with permission.)

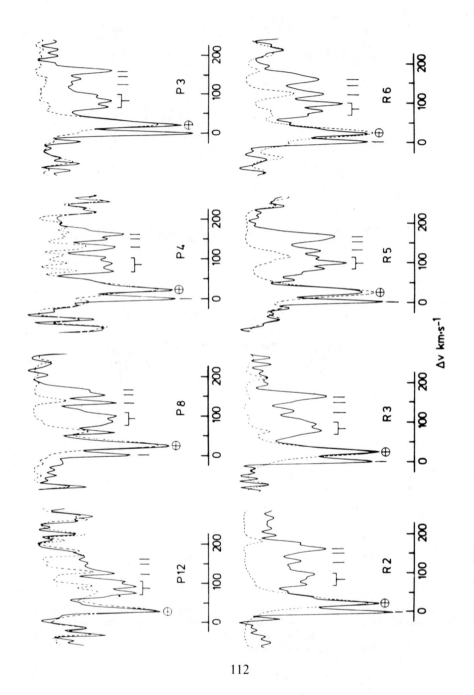

the first spaceflight of the compensated type of FTS. ATMOS operates in the "solar occultation" mode, being able to make observations twice in every orbit as the spacecraft enters and emerges from the Earth's shadow (roughly every 45 min). Such occultations occur very rapidly: an entire sequence from the time the sun is acquired well above any residual atmosphere to loss of contact behind the limb takes about 3 min. The line of sight from spacecraft to sun traverses almost vertically through the atmosphere at about $2 \text{ km} \cdot \text{s}^{-1}$. Since the required height resolution is 4 km, it follows that an ATMOS scan must be completed in 2 s. This places substantial demands on the instrument control and data systems.

ATMOS is a completely self-contained instrument in that the sun tracker used to feed sunlight into the instrument is an integral part of the package. Although its flies on a manned spacecraft, the astronauts have no involvement in its operations. The Principal Investigator for ATMOS was C. B. Farmer and for subsequent flights will be M. R. Gunson, both of JPL.

5.3.1. Purpose

ATMOS is primarily intended for global investigations of chemical abundances between 10 and 150 km, with particular emphasis on species involved in the stratospheric ozone problem. However, because of the large height coverage, a number of other problems of the upper atmosphere can also be addressed, including problems of non-LTE above 80 km (Farmer, 1987).

Limb-sounding remote sensors can only retrieve profiles for altitudes lower than themselves. The ATMOS orbit is about 350 km, so the entire atmosphere can be reached. Balloon systems are limited to altitudes below 40 km, and aircraft even lower. The basic retrieval approach is called "onion peeling" because the atmosphere is envisaged as a set of concentric shells: starting at the highest accessible layer, an estimate is made of the total abundance (column density) above that layer; to this is added the next lower shell, and so on down to (near) the surface.

5.3.2. Specifications

The specifications for the ATMOS FTS are shown in Table 5.3.

◄ ─────────────────────────────────────

Figure 5.8. Segments of a spectrum of the embedded (invisible) infrared star M8E-IR ($\delta \bar{v} = 0.06 \text{ cm}^{-1}$), showing the presence of several high velocity outflows (up to $160 \text{ km} \cdot \text{s}^{-1}$) of CO. (From Mitchell et al., 1988; reprinted with permission.)

Table 5.3. Specifications for the *ATMOS* FTS

Configuration	Compensated (cat's-eye) type; continuous scan
Spectral coverage	600–4800 cm^{-1}
Max. resolution	0.013 cm^{-1} (unapodized)
Max. OPD	± 48 cm
Scan time	2.2 s
Detector	HgCdTe at 77 K
Sensitivity	SNR > 100:1 at 2500 cm^{-1} (source-noise limited)
Operating temperature	$-5\,°C$ to $+45\,°C$
Beamsplitter and compensator	Potassium bromide
Beam diameter (internal)	2.5 cm
Telescope diameter	7.5 cm
IFOV (external)	Selectable: 1, 2, or 4 mrad
System étendue	$\geqslant 1.39 \times 10^{-4}$ cm^2·sr
Path difference control	Stabilized He/Ne laser
Weight	250 kg
Power	360 W
Data rate	16 Mbps^a

^{a}Mbps = megabits per second.

5.3.3.   Optical Layout

The optical layout for ATMOS is shown in Fig. 5.9 and the packaging in Fig. 5.10.

5.3.4.   Discussion

Some examples of ATMOS data from its first flight are shown in Figs. 5.11 and 5.12. ATMOS is scheduled for reflight as part of the *Atlas 1* mission in 1992 and at roughly 18-month intervals thereafter for a full solar semicycle of 11 years.

5.4.   THE MARK IV BALLOON/AIRCRAFT FTS

The JPL Mark IV FTS has many similarities to ATMOS but is used from the ground, aircraft, and balloons. In this way it serves to enhance and provide

▶

Figure 5.9. The ATMOS optical configuration (From Farmer, 1987. Copyright Springer-Verlag, Vienna; reprinted with permission.)

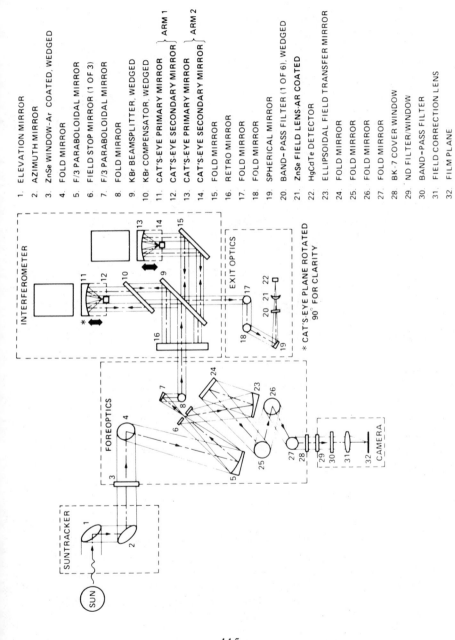

1. ELEVATION MIRROR
2. AZIMUTH MIRROR
3. ZnSe WINDOW–Ar COATED, WEDGED
4. FOLD MIRROR
5. F/3 PARABOLOIDAL MIRROR
6. FIELD STOP MIRROR (1 OF 3)
7. F/3 PARABOLOIDAL MIRROR
8. FOLD MIRROR
9. KBr BEAMSPLITTER, WEDGED
10. KBr COMPENSATOR, WEDGED
11. CAT'S-EYE PRIMARY MIRROR } ARM 1
12. CAT'S-EYE SECONDARY MIRROR
13. CAT'S-EYE PRIMARY MIRROR } ARM 2
14. CAT'S-EYE SECONDARY MIRROR
15. FOLD MIRROR
16. RETRO MIRROR
17. FOLD MIRROR
18. FOLD MIRROR
19. SPHERICAL MIRROR
20. BAND–PASS FILTER (1 OF 6), WEDGED
21. ZnSe FIELD LENS–AR COATED
22. HgCdTe DETECTOR
23. ELLIPSOIDAL FIELD TRANSFER MIRROR
24. FOLD MIRROR
25. FOLD MIRROR
26. FOLD MIRROR
27. FOLD MIRROR
28. BK–7 COVER WINDOW
29. ND FILTER/WINDOW
30. BAND–PASS FILTER
31. FIELD CORRECTION LENS
32. FILM PLANE

115

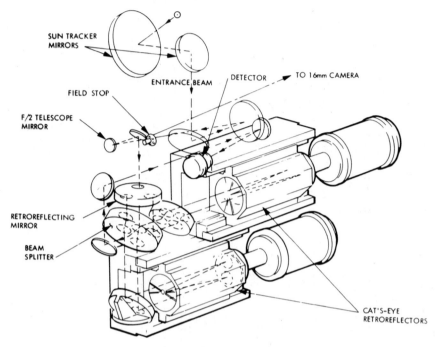

Figure 5.10. Drawing of the ATMOS flight package. (Courtesy of C. B. Farmer.)

continuity for the ATMOS data sets, which can be acquired only intermittently. Technologically, it has some advances over ATMOS, featuring double the path difference and using two detectors instead of one for optimization of sensitivity and a doubling of efficiency. The Principal Investigator is G. C. Toon of JPL (see Toon et al., 1989).

5.4.1.   Purpose

The Mark IV FTS is primarily devoted to investigations of the stratospheric ozone problem. It has been used on the ground in Antarctica and from aircraft over both polar regions to study the "ozone hole" phenomenon. Like ATMOS, the Mark IV measures solar absorption spectra from which vertical concentration profiles of most of the molecules implicated in the ozone problem can be retrieved.

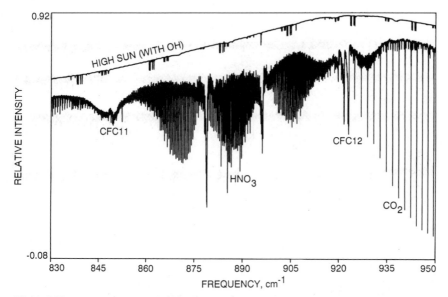

Figure 5.11. ATMOS spectra acquired well above the atmosphere (*upper curve*) and at a tangent height of 17 km (*lower curve*). Notable in the atmospheric data are features due to the two major chlorofluorocarbons CFC 11 and CFC 12, some strong features due to nitric acid vapor, and lines of background CO_2.

5.4.2. Specifications

The specifications for the Mark IV FTS are shown in Table 5.4.

5.4.3. Optical Layout

The optical layout is shown in Fig. 5.13. The packaging is essentially as drawn except that the sun tracker (an automatic device to reflect sunlight into the instrument) is on an external mount.

5.4.4. Discussion

The Mark IV scans at a relatively slow speed, so it was possible to employ a long precision lead screw for the drive system. Pitch errors in the screw, which are a serious source of periodic error, are reduced by the use of a compliant "integrating" nut that averages more than 100 turns of the 40-tpi thread. Such

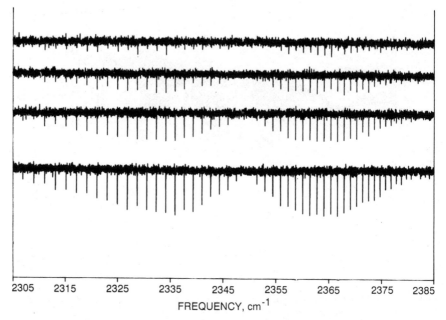

Figure 5.12. A sequence of ATMOS spectra in the Earth's upper atmosphere showing the rapid decline of CO_2 with altitude. The tangent heights (*top to bottom*) are 130, 126, 122, and 118 km.

Table 5.4. Specifications for the Mark IV Balloon/Aircraft FTS

Configuration	Compensated (cube-cornter) type; continuous scan
Spectral coverage	650–5500 cm^{-1} (1.8–15.4 μm)
Max. resolution	0.005 cm^{-1} (unapodized)
OPD	130 cm (single-sided) (200 cm max)
Scan time	210 s
Detector	InSb (1800–4100 cm^{-1}); HgCdTe (650–1850 cm^{-1}); 77 K
Sensitivity	SNR > 600 everywhere
Operating temperature	25 °C
Beamsplitter	Potassium bromide; wedged 21 arc minutes
Beam diameter (internal)	25 mm
Telescope diameter	25 mm
IFOV (external)	3 mrad
System étendue	3.6×10^{-5} cm^2·sr
Path difference control	Stabilized He/Ne
Weight	200 kg
Power	350 W
Data rate	360 kbps

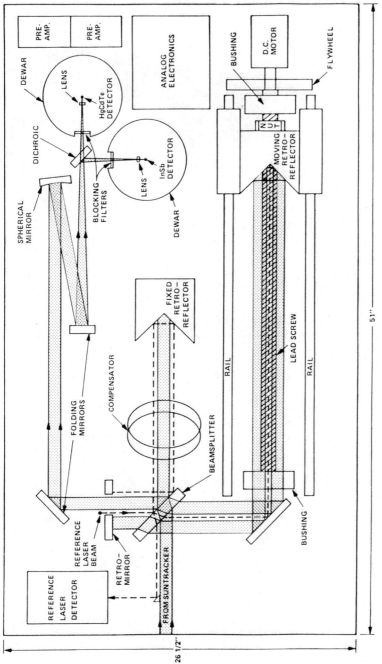

Figure 5.13. Optical configuration of the Mark IV FTS. (Courtesy of G. C. Toon.)

119

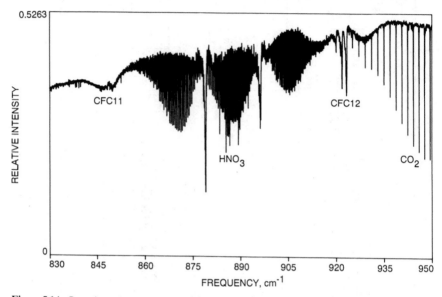

Figure 5.14. Overview of a Mark IV FTS spectrum acquired from an aircraft looking upward. The spectral coverage is identical to the ATMOS spectrum of Fig. 5.11, with which it may be compared.

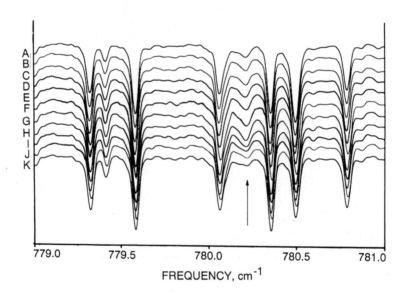

an approach was deemed essential in view of the long travel (1 m) demanded of the system (which drives only one arm) and the severe environments in which it must work. Also of note is the use of cube-corner retroreflectors, the first JPL system so equipped. Some examples of Mark IV data are shown in Figs. 5.14 and 5.15.

5.5. THE *EOS* TES FTS

The tropospheric emission spectrometer (TES) is an advanced system currently in definition for the proposed Earth Observing System (*EOS*), a series of sun-synchronous polar orbiters to be launched later this decade and at intervals for the following 20 years. It features one dimension of imaging using 1×32 line arrays of detectors and is intended for observations of Earth's

Table 5.5. Specifications for the TES FTS

Configuration	Connes-type (cube-corner); continuous scan
Spectral coverage	600–4350 cm^{-1} (2.3–16.7 μm)
Max. resolution	0.025 cm^{-1} (apodized)
Max. OPD	± 35 cm
Scan time	8 s
Detector	InSb PV (1800–4350 cm^{-1});
	HgCdTe PV (1100–1950 cm^{-1});
	HgCdTe PV (820–1200 cm^{-1});
	HgCdTe PC (600–900 cm^{-1}); all 65 K
Sensitivity (NESR at 1000 cm^{-1})	2×10^{-8} W·cm^{-1}·sr^{-1}·(cm^{-1})$^{-1}$ (source-noise limited)
Operating temperature	150 K
Beamsplitter	Potassium bromide
Beam diameter (internal)	5 cm
Telescope diameter	Selectable: 5.0 or 0.5 cm
IFOV (external)	7.5×0.75 mrad $\times$ 32 pixels or 75×7.5 mrad $\times$ 32 pixels
System étendue	9.45×10^{-5} cm^2·sr
Path difference control	Stabilized Nd:YAG laser
Weight	333 kg
Power	600 W
Data rate	15 Mbps (peak); 6 Mbps (average)

Figure 5.15. Expanded view of a limb sequence of Mark IV spectra indicating the height variability of chlorine nitrate (*arrow*), a species involved in the stratospheric ozone destruction problem. (Courtesy of G. C. Toon.)

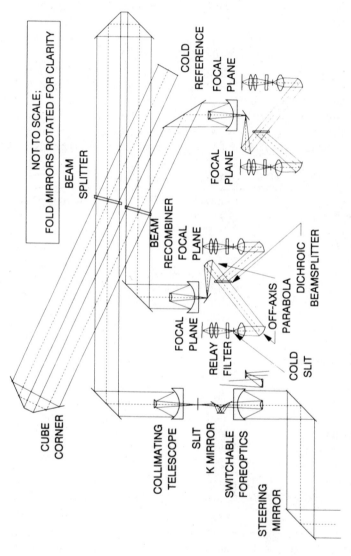

NOT TO SCALE;
FOLD MIRRORS ROTATED FOR CLARITY

BEAM
SPLITTER

COLD
REFERENCE

FOCAL
PLANE

FOCAL
PLANE

CUBE
CORNER

BEAM
RECOMBINER

FOCAL
PLANE

FOCAL
PLANE

OFF-AXIS
PARABOLA

DICHROIC
BEAMSPLITTER

RELAY
FILTER

COLD
SLIT

COLLIMATING
TELESCOPE

SLIT
K MIRROR

SWITCHABLE
FOREOPTICS

STEERING
MIRROR

Figure 5.16. Optical layout of the proposed TES FTS.

lower atmosphere (the *troposphere*) using both limb and nadir (downlooking) sounding. Unlike ATMOS and the Mark IV FTS, TES will operate purely by thermal emission and, at short wavelengths in its downlooking mode, by solar reflection. Thus, at the longer wavelengths TES will have a day/night capability, opening up the opportunity for truly global remote sensing. The Principal Investigator for TES is the present author (see Beer and Glavich, 1989).

5.5.1. Purpose

TES is intended for the investigation of problems such as global "greenhouse" warming, the gaseous precursors to acid deposition, atmospheric pollution, and gaseous exchange between the troposphere, the surface, and the lower stratosphere. In addition, TES will measure surface and atmospheric temperatures and provide information on surface material properties.

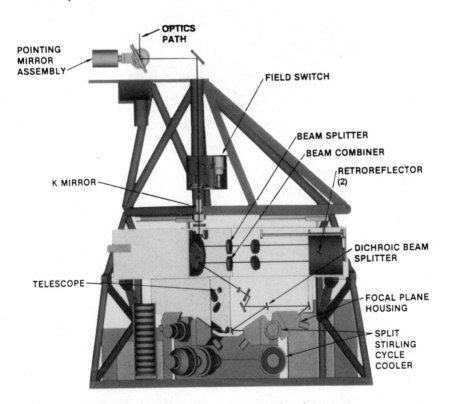

Figure 5.17. Interior view of the proposed TES instrument.

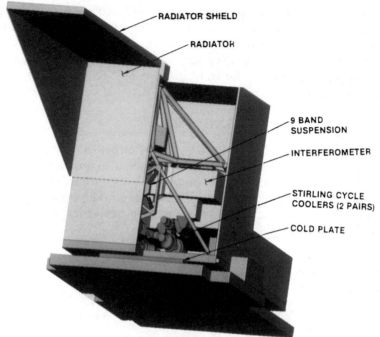

Figure 5.18. Exterior view (partial cutaway) of the proposed TES instrument (thermal layout).

5.5.2.   Specifications

The specifications for the TES FTS are shown in Table 5.5.

5.5.3.   Optical Layout

The optical layout is shown in Fig. 5.16, and the packaging in Figs. 5.17 and 5.18.

5.5.4.   Discussion

TES represents the first attempt at a new generation of spaceborne remote sensors. First and foremost, it is an imaging system, using multiple detectors

▶

Figure 5.19. Examples of downlooking and limb data from the SIRIS balloon-borne FTS. Similar data will be produced by TES on a global scale every 28 days for up to 15 years. (From Kunde et al., 1987. Copyright Optical Society of America; reprinted with permission.)

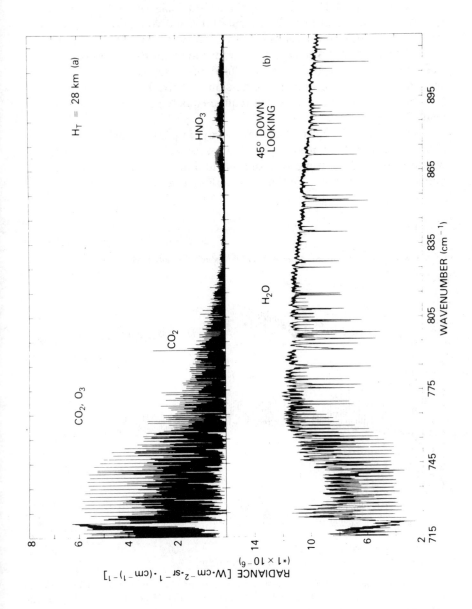

to generate simultaneous spectra at multiple locations (up to 32). As we saw in Chapter 2, an FTS can support full 2-D imagery while generating a spectrum at each image point. The data rates and volumes can, as a result, become very large. Thus, for TES, it was decided to limit the system to one dimension. Even so, TES can easily generate more than 70,000 spectra per day. For comparison purposes, the first flight of ATMOS generated some 2000 spectra over the course of its entire mission. Clearly, the impact of these new high-rate and volume systems will require equally new approaches to data analysis.

TES also uses four different detector arrays in parallel, each optimized for a different spectral region. Thus, while TES uses a Connes-type configuration with the potential for using both outputs to double the signal, it has instead been chosen to split the outputs among the four arrays. While this loses optical

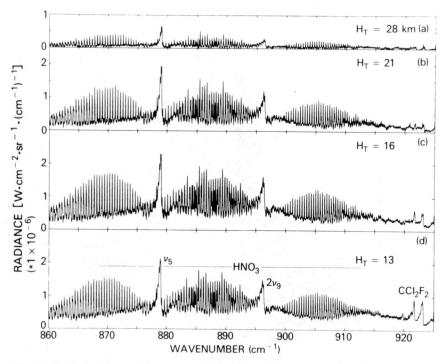

Figure 5.20. Example of a limb emission sequence also produced by the SIRIS FTS showing strong features of HNO_3. (from Kunde et al., 1987. Copyright Optical Society of America; reprinted with permission.)

efficiency, the ability to monitor four spectral regions simultaneously more than makes up for the losses.

Because TES operates by thermal emission, it will be necessary both to exercise careful control over instrument emittance and to cool the entire optical train to about 150 K after the entrance window. The external gimballed mirror provides both image-motion compensation and also permits precision pointing anywhere limb to limb. However, a two-axis gimbal system induces rotation into the FOV, so it is necessary to follow it by an "image derotator" (the Dove or K-mirror assembly) to keep the field properly aligned.

Since TES does not yet exist, no spectra can be shown. However, an example of the type of output that will be seen is shown in Figs. 5.19 and 5.20. The spectra were generated by a balloon-borne system called SIRIS (Stratospheric IRIS) from the Goddard Spaceflight Center. Note that the limb-viewing spectral lines are now maxima, rather than the minima seen in most previous examples.

CHAPTER

6

REMOTE SENSING ENVIRONMENTS

Since this book is concerned with the utilization of FTSs in remote sensing, it is appropriate to add a few words about the environments in which such systems must survive and operate. Generally speaking, these are far from benign and a typical commercial system, for example, would need substantial modification before it was operable at a telescope, balloon, or aircraft. While such modified laboratory systems have indeed been successfully used, many (if not most) are purpose-built instruments and frequently "one off" handcrafted gems. This is almost universally true of rocket- and spacecraft-based systems, where the added safety and reliability requirements add greatly to the costs.

6.1. TELESCOPE-BASED SYSTEMS

Beginning closest to the Earth's surface, we will look at the environment of a typical astronomical observatory. Most telescopes offer several mounting positions for equipment, but the size and complexity of a typical FTS is such that only those capable of withstanding significant loads and providing adequate logistics (power, vacuum lines, high-speed data links, cryogen supplies, etc.) are normally used. The simplest but least efficient is a *coudé* (fixed) focus that pipes the light into an environment that is typically somewhat laboratory-like, including air conditioning. The low efficiency results from the large number of reflections usually needed to get to the coudé room—at least four and frequently as many as six or seven. Furthermore, the available FOV at a typical coudé focus is quite small, a few degrees at best, and the image quality is frequently rather poor owing to the long air path within the observatory building. As we have noted in earlier chapters, image quality is important because one wants to use the smallest possible field stop to reduce unwanted background. However, as suggested above, large pieces of hardware can be assembled in a coudé room and operated under relatively stable conditions. Indeed, in a bygone era, coudé grating spectrographs achieved quite gargantuan proportions and weights of many tons.

Most of the next generation of large telescopes will use elevation-azimuth (el-az) mounts, which are more compact and stable than the traditional polar mount. An additional benefit is that, with only three reflections, a so-called Nasmyth focus is available on the axis of the elevation bearing. A platform attached to the azimuth bearing rotates with the telescope but maintains a fixed gravity vector, making it very attractive for large instruments such as an FTS. Furthermore, the image quality and FOV should be markedly better than at a coudé focus.

Best efficiency, image quality, and FOV is obtained at the Cassegrain focus just behind the telescope primary mirror after only two reflections. It is also the most difficult to use (from an FTS standpoint) because the gravity vector changes continuously. This complicates instrument design, exacerbated by the constricted available space and limited clearances owing to the telescope structure. It was therefore many years after the first coudé FTS systems were deployed that the first truly successful Cassegrain FTS (see Chapter 5) could be implemented.

Common to all astronomical instruments are some serious environmental problems that must be overcome. First, observatories are (almost) always built in high, dry locations. A good electrical ground is hard to come by. In one instance of which I have personal knowledge (and the blackened ruin of a digital control system to prove it), a so-called ground bus was found, in fact, to have 20-Ω resistance between it and a true ground. A lightning strike on the busbar revealed the truth in a dramatic fashion. You have been warned!

The remoteness of most observatories also makes their electrical power of dubious quality. If they are connected to a grid, they are usually the last link in a very long chain. If not, local generators are used. In either case, quite large voltage fluctuations and non-sinusoidal waveforms are commonplace, as are frequent losses of all power (usually in the middle of a critical observation, of course). The wise designer therefore incorporates uninteruptible power supplies, surge protectors, and line isolators at the outset.

There are other consequences of operating in an exposed location at high altitudes:

- High winds can bring dust storms that have an insidious habit of penetrating even the most tightly enclosed optics and mechanisms (if light can get in, so can dust).
- Low atmospheric pressure can cause problems with vacuum-column-tensioned tape transports and Winchester disks. It also complicates the cooling of electronic racks (double up on the fans!).

- High ambient ozone concentrations cause havoc with conventional insulators [neoprene, poly(vinyl chloride)]. Teflon® holds up much better.
- Altitude sickness results not just in nausea and headache but also a strong tendency to make silly (even dangerous) mistakes. Make interactive controls as "user friendly" as possible.

Many of these strictures are equally applicable to aircraft and balloon systems, which we consider next.

6.2. AIRCRAFT-BASED SYSTEMS

Of all possible remote sensing platforms, the environment of an aircraft is undoubtedly the worst. Engine vibration, turbulence, and boundary-layer effects all create severe problems over and above the obvious one of accommodation. Decoupling an FTS from the vibration (or making it so rugged that it is irrelevant) is nontrivial. Furthermore, many aircraft offer only 28-V DC power. If AC power is available, it is usually 400 Hz, so special transformers and motors (for example) are required.

Research aircraft come in all sizes from the "high and fast" NASA ER-2 (which requires totally autonomous instrumentation), through the "flying laboratories" such as the NASA DC-8 (which permits "hands on" operation by teams of investigators), to the "low and slow" such as an Electra or a helicopter (but much less luxurious than a DC-8!). Nevertheless, FTSs have operated successfully in all these environments.

The main recommendation is to engage the services of an aerodynamicist before you start. Quite apart from the safety issues of cutting observing ports (with or without windows) into the skin of an aircraft, the buffeting and microturbulence can cause serious problems to instruments requiring high-quality imaging. Boundary-layer fences are often essential on the aircraft exterior, and their design and installation is well beyond the capability of most instrumentalists.

Designers of open-ported (i.e., aircraft ambient temperature and pressure) instruments must also be alert to the dangers of condensation forming on exposed optical surfaces during the descent. At least one instrument of which I am aware (not an FTS) required a major redesign for this very reason.

6.3. BALLOON-BORNE SYSTEMS

In this category I include not just the familiar free-flying balloons but also the less obvious blimps and aerostats (tethered balloons). These latter, while limited to quite low altitudes (1500–3000 m) can provide a quiet and stable environment for quite surprisingly large payloads. Aerostats have the added advantage that ground power can be "piped" up to the instrumentation. The pilots of blimps are very helpful and are quite used to strange requests from observers because one of their major activities is to accommodate TV cameras at major sporting events. Furthermore, they can almost (not quite) hover at any altitude.

Free-flying balloons come in sizes varying from an "little" as 30,000 m³ all the way up to monsters of 10^6 m³. Most can lift payloads of 2000 kg or more, the difference being the attainable float altitude. The smallest are limited to 20–25 km; the biggest can approach 40 km. Endurances of 24 h or more are common, the limitation usually being not the balloon itself but the range of the telemetry (a few hundred kilometers unless a mobile receiver is available).

Stratospheric winds can be very high (200 km/hr not being unknown) except at times during two seasons of the year, spring and fall, when the winds reverse direction. These "turnaround" times are therefore the most favored period (a few weeks) for launching long-duration missions. Thus balloon-launch facilities are very active at these times, and long queues for flights are common. Furthermore, the logistics required for a balloon flight are substantial, so the number of possible launch sites in the world is relatively small. Experiments needing to overfly a specific location (i.e., downlooking sensors) will therefore have great difficulty in accomplishing their goals (believe me, I know).

Scientific ballooning has achieved great sophistication in recent years, primarily as a result of the need for measurements to study the stratospheric ozone problem. Almost any kind of experiment (*in situ*, active, or passive) can be accommodated under conditions that are generally within the capability of quite modest research groups. Payload safety is, of course, a concern. No one wants a piece of scientific hardware falling uncontrolled out of the sky (it has happened!). Furthermore, the balloon environment, though reasonably quiet, is not totally benign. Launch and landing shocks reach several *g*'s, and equipment cooling under intense isolation is difficult. High voltages are also a problem: stratospheric pressures are ideally constituted to promote arc-over in power supplies, which must therefore either be evacuated or pressurized. Power consumption must be carefully controlled not only because of the cooling problem but also because batteries, the only source of electrical power, are heavy and expensive (albeit rechargeable between flights).

6.4. SPACECRAFT SYSTEMS

The primary difference between spaceborne systems and any of the foregoing is the matter of recovery and reflight. Except for experiments carried on the space shuttle and a few highly publicized astronaut repairs, the fundamental requirement is that the instruments work first time, every time, and keep on working for many years. The fact that some do not does not alter the requirement. "Test flights" such as are commonly done with balloons and aircraft are essentially unknown in space science.

The consequences of this emphasis on reliability are dramatically increased costs and development times. Costs one to two orders of magnitude greater than those of balloon or aircraft systems are not unusual, not are development times of a decade or more. Nevertheless, for certain types of measurement, especially those requiring a global viewpoint or a close look at a planet, there is no substitute.

Space hardware development being such a specialized business, there is little useful that one can say on the subject. Anyone planning on proposing a space experiment will necessarily become allied with specialists and learn more than he or she ever wanted to know about this difficult but perennially exciting field. Countless acronyms are the order of the day, as are the almost incredible volumes of documentation and interminable meetings that accompany the exercise.

The space environment is extremely variable. In near-Earth orbit, the residual atmosphere is a constant source of concern because its high reactivity plays havoc with exposed materials (not to mention the hazard of assorted space debris). Contamination from outgassing and thruster firings is also a problem. This is especially true for cooled instruments that act as cryopumps for anything condensible. However, even uncooled instruments are not immune. Careful protection and provision for self-cleaning is essential.

Almost anywhere in space, so-called *single event upsets* (SEU) in electronics caused by charged particle radiation are hazards. One consequence of this is that the need to use radiation-hardened parts results in electronic systems that are surprisingly primitive by ground-based standards. For example, the personal computer on which this book is being written has far greater computing power than anything that has ever flown on an unmanned spacecraft.

One final comment. A pleasant surprise of the space environment is that mechanisms (e.g., the moving arms of an FTS) often work better in space than they do in the lab—no gravity-induced friction.

CHAPTER

7

GENERAL OBSERVATIONS AND CONCLUSIONS

Through the previous six chapters we have seen how FTSs operate; how they can be used as remote sensors; and some of the pitfalls that lie in wait for the unwary. As we noted at the outset, an FTS can be almost frighteningly efficient at data collection but that its implementation is by no means simple. In such a brief overview it has been possible only to offer some familiarity with the concept and, equally important, the jargon that has accumulated over the past 30 years. Those who have been patient enough to complete this book may be less baffled by the terminology common to this field.

Much has been left unsaid. The "gory details" of design have been ignored, yet it is in these details that success or failure is found. In particular, I have said little about path-difference control systems, the heart of an FTS, because (a) they are very specific to the system and (b) the analog controls used in the past are clearly being superseded by digital approaches, so anything said now would soon be out of date.

We have also covered some successful examples of remote sensing FTSs; the failures could fill their own book.

Only two final points remain to be made:

- Optical design generally concentrates on the geometrical properties of an optical system. FTSs demand additional insight into the *physical optics* properties of the system. Indeed, these are generally the more important considerations. Just because a system produces superb spot diagrams will not make it an FTS.

- The mechanical design of an FTS must take into account effects that are orders of magnitude below those of conventional optical systems. On the scale of an FTS *the world is made of rubber!*

OPTIMUM FILTERS

If we define *discrimination* as an appropriate balance between generating an output spectrum that is a faithful rendition of the input while preserving the best possible signal-to-noise ratio (SNR), it seems intuitively obvious that the spectral resolution should match the width of the lines in the spectrum. However, in order to decide how close the match must be, it is necessary to explore the mathematics of the process. This is done through the *optimum filter theorem*.

A.1. THE OPTIMUM FILTER THEOREM

Imagine a linear time-invariant filter whose response to a unit impulse $\delta(t)$ is the *impulse response* $a(t)$. The *transfer function* of the filter is defined as the Fourier transform of $a(t)$:

$$A(f) = \int_{-\infty}^{\infty} a(t) \exp(2\pi i f t)\, dt \qquad (A.1)$$

The response of the filter to an arbitrary input $x(t)$ is the convolution of the input with the filter impulse response:

$$y(t) = \int_{-\infty}^{\infty} x(\tau) \cdot a(t - \tau)\, d\tau \qquad (A.2)$$

and if the Fourier transform of $x(t)$ is $X(f)$, then, by definition, the output $Y(f)$ is

$$Y(f) = X(f) \cdot A(f) \qquad (A.3)$$

Thus we can rewrite Eq. A.2 as

$$y(t) = \int_{-\infty}^{\infty} X(f) \cdot A(f) \exp(2\pi i f t) \, df \qquad \text{(A.4)}$$

The *power density spectrum* of $Y(f)$ is defined as

$$|Y(f)|^2 = |X(f)|^2 \cdot |A(f)|^2 \qquad \text{(A.5)}$$

Now, the Weiner theorem for the autocorrelation $\Phi_{11}(\tau)$ of a random function $n(t)$ states that it is related to the power density spectrum $|N(f)|^2$ via a standard cosine Fourier transform:

$$|N(f)|^2 = \int_{-\infty}^{\infty} \Phi_{11}(\tau) \cos(2\pi f \tau) \, d\tau \qquad \text{(A.6)}$$

where

$$\Phi_{11}(\tau) = \int_{-\infty}^{\infty} n(t) \cdot n(t + \tau) \, dt \qquad \text{(A.6a)}$$

Thus it is legitimate to write

$$|N_y(f)|^2 = |N_x(f)|^2 \cdot |A(f)|^2 \qquad \text{(A.7)}$$

where $|N_x(f)|^2$ represents the input noise power density spectrum and $|N_y(f)|^2$ the output.

The variance of the output noise $\langle n^2(t) \rangle$ is the autocorrelation of $N_y(f)$ at zero lag:

$$\langle n^2(t) \rangle = \int_{-\infty}^{\infty} |N_y(f)|^2 \, df \qquad \text{(A.8)}$$

$$= \int_{-\infty}^{\infty} |N_x(f)|^2 \cdot |A(f)|^2 \, df \qquad \text{(A.8a)}$$

Thus the expected square of the output SNR given an input $x(t)$ and noise

$n(t)$ is

$$\frac{|y(t)|^2}{\langle n^2(t)\rangle} = \frac{\left|\int_{-\infty}^{\infty} X(f)\cdot A(f)\exp(2\pi i f t)\,df\right|^2}{\int_{-\infty}^{\infty} |N_x(f)|^2\cdot|A(f)|^2\,df} \tag{A.9}$$

and the problem devolves to one of finding the maximum value for Eq. A.9.

Lemma (After Lee, 1960). *That the sum of the autocorrelations of any two functions evaluated at zero lag is equal to or greater than twice the absolute value of their cross-correlation at any lead or lag.*

Consider the integral

$$\int_{-\infty}^{\infty} [f_1(t) \pm f_2(t + \tau)]^2\,dt$$

Provided that $f_1(t) \neq f_2(t + \tau)$ at all τ, the integral is necessarily positive. Thus, after expansion,

$$\int_{-\infty}^{\infty} f_1^2(t)\,dt + \int_{-\infty}^{\infty} f_2^2(t + \tau)\,dt \pm 2\cdot\int_{-\infty}^{\infty} f_1(t)\cdot f_2(t + \tau)]\,dt \geqslant 0$$

But

$$\int_{-\infty}^{\infty} f_n^2(t)\,dt = \Phi_{nn}(0); \qquad n = 1, 2$$

the autocorrelation at zero lag, and

$$\int_{-\infty}^{\infty} f_1(t)\cdot f_2(t + \tau)]\,dt = \Phi_{12}(\tau)$$

the cross-correlation at lag τ. Hence

$$\Phi_{11}(\tau) + \Phi_{22}(\tau) \geqslant \pm 2\Phi_{12}(\tau); \qquad -\infty < \tau < \infty$$

which is equivalent to

$$\Phi_{11}(\tau) + \Phi_{22}(\tau) \geqslant 2 \cdot |\Phi_{12}(\tau)|; \qquad -\infty < \tau < \infty$$

and the lemma is demonstrated.

Referring back to Eq. A.9, the top line is recognizable as the cross-correlation of $X(f)$ and $A(f)$ and, from the foregoing lemma, that

$$\int_{-\infty}^{\infty} X^2(f)\,df + \int_{-\infty}^{\infty} A^2(f)\,df \geqslant 2 \cdot \left| \int_{-\infty}^{\infty} X(f) \cdot A(f) \exp(2\pi i f t)\,df \right|$$

for all t.

Thus we may immediately infer that the top line of Eq. A.9 will reach its maximum value when $A(f) \equiv X(f)$. That is, the maximum signal power (and hence information) is transmitted when the filter response matches the spectrum of the input signal. Provided that the noise power density spectrum is reasonably well behaved [$N(f)$ being constant or declining with increasing frequency f], the same condition also ensures that the maximum possible SNR will be preserved. Thus the optimum filter theorem is proven.

A.2. IMPLEMENTATION

The theorem cannot be used in a simpleminded fashion. Were we to have perfect knowledge of the spectral characteristics of the signal in advance (in order to construct our optimum filter by whatever means), there would be little point in making the observation. This is in accord with one of the fundamental precepts of information theory that describes the information content of a message (the spectrum in our case) as being proportional to minus the logarithm of its probability. That is, a message whose content is certain ($P = 1$) contains no (new) information. Note that, for a well-built FTS, the noise properties are generally well behaved (i.e., near gaussian) and can usually be inferred from the data themselves (Beer and Norton, 1988).

We must therefore content ourselves with creating a filter based on fore-knowledge of the general characteristics of the signal, in turn derived from the physics and chemistry outlined in Chapter 3. The resultant filter will not be "optimum" in the sense of the theorem but will be reasonably close. In any case, the loss of signal discrimination through the use of nonoptimum

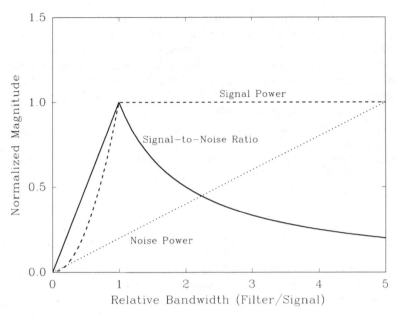

Figure A.1. Graphic illustration of Eq. A.9 using a simple boxcar spectrum with added gaussian noise passing through a boxcar filter: *dashed line*, numerator; *dotted line*, denominator; *solid line*, ratio.

filtering can be kept quite small: Fig. A.1 shows how discrimination varies for $A(f) \not\equiv X(f)$ for a particular (and simple) test case.

The case assumed is one of a signal having a rectangular spectrum— amplitude being constant between frequencies f_1 and f_2, and 0 elsewhere. Likewise, the filter is rectangular between f_1 and some adjustable frequency f_3. The input noise, on the other hand, is assumed to be "white"—mean square value being constant between 0 and ∞ (or, in practicality, some frequency $\gg f_2$ or f_3). As the filter bandpass increases from zero, the signal *power* increases quadratically until all is passed, whereafter it remains constant. The noise power increases linearly with bandwidth. Consequently, the ratio shows a maximum when $f_2 = f_3$ as the theorem predicts because as the bandwidth of the filter increases beyond f_2, only noise is added. Were the filter bandwidth to increase indefinitely, the noise power would eventually overwhelm the signal. However, even for this quite extreme example, essentially 50% efficiency or better is achieved for a factor of 2 change in filter band-pass. Less abrupt filters would show a less sharply peaked profile.

Returning now to the real world of spectrometry, we have already seen that the widths of isolated spectral features are reasonably constant on a frequency scale. Furthermore, while atmospheric spectra are very dense, they are not *so* dense that individual lines are lost. Thus a spectrometer whose instrumental line shape (ILS, its impulse response, as we have previously shown in Chapter 2) is essentially the same as that of the lines to be observed will constitute a good approximation to an optimum filter.

It is, of course, implicit in the theorem that the ILS should be applied to all spectral lines simultaneously. This is inherent in the FTS and dispersive array spectrometers; we have already seen (Chapter 4) how sequential scanning of a spectrum (i.e., essentially filtering the spectrum down to a single line) entails a massive loss in efficiency.

A.3.   AN ALTERNATIVE APPROACH

For completeness, we must briefly discuss an alternative approach that, too, has found an "ecological niche" in remote sensing: the *gas correlation spectrometer* (GCS). The basis of the GCS is the recognition that the near-ideal filter for the signal from a particular molecule is the molecule itself. In the simplest implementation, the scene is viewed alternately through a cell containing the gas of interest and an identical, evacuated cell. The filter is thus a "negative" one in that the gas cell blocks out, rather than transmits, the radiation of interest. Alternative (and widely used) implementations modulate the amount of gas in the cell periodically [typically at a few tens of hertz (Houghton et al. 1984)]. The GCS has much to commend it: it is relatively simple, achieves remarkable discrimination even in the face of severe interference from other species, and—because it measures many lines simultaneously—achieves an excellent SNR. The defects are as follows: (1) The gas of interest must be containable, which eliminates all free radicals (e.g., OH) and highly reactive species (e.g., O_3) from contention; even species such as HCl and HF require special techniques to contain them when lifetimes of years are required. (2) The physical state of the gas must be not too far removed from that of the source, which makes it useful for Earth and planetary atmospheres but not, for example, the interstellar medium wherein the very character of the spectrum (CO, for example) is quite unlike that produced at Earth ambient conditions. (3) Since the GCS is a local (and fixed) frequency standard, it follows that Doppler shifts between source and GCS can introduce spurious decorrelation. Thus the GCS must, on an orbiting or flyby spacecraft, always observe in a direction perpendicular to the velocity vector. A more advanced (but developmental) approach interposes an electro-optical phase/frequency modulator into the incoming beam. In this way, small Doppler shifts can be compensated and, for example, used to measure winds along the line of sight.

Beer and Norton (1988) also showed that it is possible to apply this correlation concept to FTS spectra after the fact, using a computer-generated model of the spectrum as the filter. By this means, they were able to extract useful information about spectral features in the Earth's upper atmosphere that were (apparently) totally buried in noise. They also showed that the same technique inherently provides an elegant means of extracting accurate Doppler shifts from the data.

BIBLIOGRAPHY AND REFERENCES

REFERENCES FOR CHAPTER 1

Bingham, E. O. (1974). *The Fast Fourier Transform*. Prentice-Hall, Englewood Cliffs, New Jersey.

Bracewell, R. N. (1990). "Numerical Transforms." *Science* **248**, 697–704.

Klein, M. V., and Furtak, T. E. (1986). *Optics*, 2nd ed. John Wiley & Sons, New York.

Press, W. H., Flannery, B. P., Teulosky, S. A., and Vetterling, W. T. (1986). *Numerical Recipes*. Cambridge University Press, London and New York.

Smithsonian Meteorological Tables, 6th ed. (1968). Smithsonian Institution, Washington, D.C.

REFERENCES FOR CHAPTER 2

Connes, J., Delouis, H., Connes, P., Guelachvili, G., Maillard, J-P., and Michel, G. (1970). "Spectroscopie de Fourier avec Transformation d'un Million de Points." *Nouv. Rev. Opt. Appl.* **1**, 3–22.

Jacquinot, P. (1954). "The Luminosity of Spectrometers with Prisms, Gratings, or Fabry–Perot Etalons." *J. Opt. Soc. Am.* **44**, 761–765.

REFERENCES FOR CHAPTER 3

Allen, H. C., and Cross, P. C. (1963). *Molecular Vib-Rotors*. John Wiley & Sons, New York.

Goody, R. M., and Yung, Y. L. (1989). *Atmospheric Radiation—Theoretical Basis*. Oxford University Press, London and New York.

Humlicek, J. (1982). "Optimized Computation of the Voigt and Complex Probability Functions." *J. Quant. Spectrosc. Radiat. Transfer* **27**, 437–444.

Rodgers, C. D. (1976). "Retrieval of Atmospheric Temperature and Composition from

Remote Measurements of Thermal Radiation." *Rev. Geophys. Space Phys.* **14**, 609–624.

U.S. Standard Atmosphere, 1976. National Oceanic and Atmospheric Administration, Washington D.C.

Wayne, R. P. (1991). *The Chemistry of Atmospheres*, 2nd ed. Oxford Univ. Press Clarendon Press, London and New York

REFERENCES FOR CHAPTER 4

Beer, R. (1967). "Fourier Spectrometry from Balloons." *Appl. Opt.* **6**, 209–212.

Beer, R., and Marjaniemi, D. (1966). "Wavefronts and Construction Tolerances for a Cat's-Eye Retroreflector." *Appl. Opt.* **5**, 1191–1197.

Breckinridge, J. B., and Schindler, R. A. (1981). "First-Order Optical Design of Fourier Spectrometers." *In Spectrometric Techniques* (Vanasse, G. A. ed.), Vol. II, pp. 63–125. Academic Press, New York.

Forman, M. L., Steel, W. H., and Vanasse, G. A. (1966). "Correction of Asymmetric Interferograms Obtained in Fourier Spectroscopy." *J. Opt. Soc. Am.* **56**, 59–63.

Guelachvili, G. (1981). "Distortions in Fourier Spectra and Diagnosis." *In Spectrometric Techniques* (Vanasse, G. A. ed.), Vol. II, pp. 1–62. Academic Press, New York.

Keyes, R. J., ed. (1977). *Optical and Infrared Detectors.* Springer-Verlag, Berlin.

Kohlenberg, A. (1953). "Exact Interpolation of Band-Limited Functions." *J. Appl. Phys.* **24**, 1432–1436.

Kruse, P. W. (1977). "The Photon Detection Process." *In Optical and Infrared Detectors* (R. J. Keyes, ed.), pp. 5–69. Springer-Verlag, Berlin.

Shannon, C. E. (1949). "Communication in the Presence of Noise." *Proc. Inst. Radio Eng.* **37**, 10–21.

Steel, W. H. (1971). "Interferometers for Fourier Spectroscopy." *In Proceedings of the Aspen International Conference on Fourier Spectroscopy, 1970*, AFCRL Spec. Rep. 114 (AFCRL-71-0019), 43–53.

Zachor, A. S., Coleman, I., and Mankin, W. G. (1981). "Effects of Drive Nonlinearities in Fourier Spectroscopy." *In Spectrometric Techniques* (Vanasse, G. A. ed.), Vol. II, pp. 127–160. Academic Press, New York.

REFERENCES FOR CHAPTER 5

Beer, R., and Glavich, T. A. (1989). "Remote Sensing of the Troposphere by Infrared Emission Spectroscopy." *Proc. Soc. Photo-Opt. Instrum. Eng.*, **1129**, 42–51.

Kunde, V. G., Brasunas, J. C., Conrath, B. J., Hanel, R. A., Herman, J. R., Jennings,

D. E., Maguire, W. C., Walser, D. W., Annen, J. N., Silverstein, M. J., Abbas, M. M., Herath, L. W., Buijs, H. L., Berube, J. N., and McKinnon, J. (1987). "Infrared Spectroscopy of the Lower Atmosphere with a Balloon-Borne Fourier Spectrometer." *Appl. Opt.* **26**, 545–553.

Farmer, C. B. (1987). "High Resolution Infrared Spectroscopy of the Sun and the Earth's Atmosphere from Space." *Mikrochim. Acta* [Wien], **111**, 189–214.

Hanel, R., Crosby, D., Herath, L., Vanous, D., Collins, D., Creswick, H., Harris, C., and Rhodes, M. (1980). "Infrared Spectrometer for *Voyager*." *Appl. Opt.* **19**, 1391–1400.

Hanel, R. A. (1981). "Fourier Spectroscopy on Planetary Missions including *Voyager*." *Proc. Soc. Photo-Opt. Instrum. Eng.*, **289**, 331–344.

Maillard, J.-P., and Michel, G. (1982). "A High-Resolution Fourier Transform Spectrometer for the Cassegrain Focus at the CFH Telescope." *In Instrumentation for Astronomy with Large Optical Telescopes* (Humphries, C. M. ed.), pp. 213–222. Kluwer Academic Publishers/Reidel Publishing Co., Norwell, Massachusetts.

Mitchell, G. F., Allen, M., Beer, R., Dekany, R., Huntress, W., and Maillard, J.-P. (1988). "The Detection of High Velocity Outflows from M8E-IR." *Astrophys. J. Lett.*, **327**, L17–L21.

Owen, T., Maillard, J.-P., de Bergh, C., and Lutz, B. L. (1988). "Deuterium on Mars: The Abundance of HDO and the Value of D/H." *Science* **240**, 1767–1770.

Toon, G. C., Farmer, C. B., Schaper, P. W., Blavier, J-F., and Lowes, L. L. (1989). "Ground-Based Infrared Measurements of Tropospheric Source Gases over Antarctica During the 1986 Austral Spring." *J. Geophys. Res.* **94**, 11613–11624.

REFERENCES FOR APPENDIX

Beer, R., and Norton, R. H. (1988). "Analysis of Spectra Using Correlation Functions." *Appl. Opt.* **27**, 1255–1261.

Houghton, J. T., Taylor, F. W. and Rodgers, C. D. (1984). *Remote Sounding of Atmospheres.* Cambridge University Press, London and New York.

Lee, Y. W. (1960). *Statistical Theory of Communication.* John Wiley & Sons, New York.

INDEX

Major text references are in **boldface**; those to figures or tables are in *italics*.

Absorption coefficient, 42, **49–50**
Accuracy, frequency, 22–24
Advantage, 18–20, 22, 65, 83, 132
  Connes, 22
  Fellgett (multiplex), 19, 20, 65
  Jacquinot, 19
  multichannel, 20
Aliasing, **70**, 77
Alias order, **70–73**, 77
Amagat, 4
Amplitude division, 6
Analog-to-digital (A–D) converter, **82**
Ångstrom, 4
Angular dispersion:
  FTS, 9, **26**
  grating spectrometer, **18**
Angular slit length, grating spectrometer, 18
$A\Omega$ product, 15–19, *16*, 61, 63, 67
Apodization, 14
ATMOS FTS, 51, 101, **109–114**, 116,
  115–116, 123, 127
Atmosphere:
  model, **34**
  U.S. Standard, **31**, *32*, 34
Atmospheric composition, 33, **35**, *35*, 51, 101
Atmospheric density, 4–5, **33**, 34, 37
Atmospheric pressure, 4, 22, 31, **33–34**, 37,
  48, 49, 51
Atmospheric scale height, **33**, 36
Atmospheric temperature, 4, 5, **31–33**, *32*,
  41–42, 46–48, 51, 101, 123
Atmospheric turbulence, **68**, 77, 131
Atmospheric window, *43*, **44**, 67

Background photon flux density (BPFD),
  65, **67–68**

Backgrounds, 29, **55**, **57**, 62, *65*, 67, 87,
  107–108, 129
Bandwidth:
  electrical, 62, 77, 91, 93, 108, 141
  spectral, **10**, 76, 141
Beamrecombiner, 7, 85, 89, 107, 108
Beam shear, **85**, 87
Beamsplitter, 6–9, 16, 17–18, 24, 25, 80, 81,
  85, 87, 89, 91, 107
  crystalline, 16, 85, 87, 89, 102, 107–108,
  114, 118, 121
  thin film, 7, 16
Bit error rate (BER), **98–99**
Boltzmann distribution, 46
Boxcar and sinc functions, 11–14, 22
Brightness temperature, **56–57**, 58, *58*, 63
Broadening:
  Doppler, **48–51**
  Lorentz, **48–51**

Calibration:
  gain, **97–98**
  radiometric, 77, 91, 93, **97–98**
  spectral, **22–24**
Canada–France–Hawaii telescope (CFHT)
  FTS, **107–109**, *109–110*
Centimeter·amagat, 4
Channeling (channel spectra), **91**, *92*
Column density, 4, 50, 113
Compensator, **80**, 89
Concentration, parts per —, 4
Connes advantage, 22
Constructive interference, 5
Contamination, 133
Contribution function, 41
Convolution, 12, 13, 39, 137

Cosine wave, 2, 7, 28, 69, 71, 89
Cross-correlation, 139–140, 143
Cryogenic postdisperser, 67

Data:
  allowable bit error rate (BER), **98–99**
  compression, 73, 96
  linearity requirements, 94, 96, **97**
  rate and volume 73, *99*, 99–100, *102, 108,
      114, 118, 121*
Density, column, 4, 50, 113
Destructive interference, 5, 7
Detectivity, **60**, *65*, 67–68
Detectors, 7–8, 13, 18–20, 22, 24, **25–29**,
    58–60, 62–63, 65–69, 74, 77–82, 84,
    86, 91, 93, 97, 116, 121, 124, 126
  infrared, **59–60**, *102, 108, 114, 118, 121*
  saturation effects, 97
  visible and UV, 59
Dewar, 81–82
Dewpoint, 4, 5
Digitization, **94–98**, *94, 95, 99*
Diffraction grating, 16, 18–19, 21, 26, 70, 129
Discrete Fourier transform, 14
Division:
  by amplitude, 6
  by wavefront, 6
Doppler broadening, 48
Doppler effect, 23, 48
Drive motors, 80–81, 131
Dynamic range, *65*, 87, **94**

Efficiency, modulation, 24, 25, 62, 85
Electromagnetic interference (EMI), **93**
Electromagnetic waves, 1
  amplitude, 1–3, 5–7, 24, 28
  intensity, 5, 7, 9–10, 24
Emittance, 37, 41–42, 56, 58, *58*, 60–61, *61*,
    63, 126
Environments, 58, 83, 101, **129–133**
  aircraft, 93, 121, **131**
  balloons, **132**
  spacecraft, **133**
  telescopes, 107, **129–131**
EOS TES FTS, **121–127**, *122–124*
Equilibrium:

hydrostatic, **33**
local thermodynamic, 33, 113
Étendue, 15–19, *16*, 61, 63, 67
  Fourier transform spectrometer (FTS), **17**
  grating spectrometer (GS), **18**
  FTS and GS compared, 18

Fast Fourier transform (FFT), 14
Fellgett advantage, 19, 65
Field of view (FOV), 130
  finite, 8–9
Filters:
  electronic, 70, 82
  optical, 15, 25, 62–63, 67, 70–71, 76, *76*,
      81–82, 91
  optimum, 51, **137–143**
Focal plane arrays, 20, 60, 142
  use in FTS, **25–28**, *27*, **78–79**, 109, 121, 126
  use in dispersive spectrometers, *28*, **29**,
      66–67
Fourier transform, 8, **10–14**
Fourier transform spectrometer (FTS), 11
  bearings for, 81, 87
  Canada–France–Hawaii telescope,
      107–109, *109–110*
  Connes-type, **85–87**, *85*, 93, 126
  drive motors for, 80–81, 131
  EOS TES, **121–127**, *122–124*
  balloon/aircraft Mark IV, **114–121**, *119*
  Michelson-type, 7, *8*, **15**, 55, *79*, 80, **83**, 102
  rapid (continuous) scanning, **78–79**, 82,
      89, *102, 114, 118, 121*
  Space Shuttle ATMOS, 51, 101, **109–114**,
      115–116, 116, 123, 127
  step-and-lock scanning, **78–79**, 91, 108, *108*
  tilt-compensated, **87–89**, *88*
  Voyager IRIS, 15, **101–104**, *103–105*
Frequency (unit), 1–2, 4
Frequency, electrical, **7**, 19, 60, 69, **72–73**,
    74, 77, 78, 81, 89–90, 93
Frequency accuracy, 22
  Fourier transform spectrometer, **23**
  grating spectrometer, **23–24**
Fringes, 23
Function:
  boxcar, 11, 13
  complex, 10

contribution, **41**,
Planck, 37, 41, 56, 63
real, 10
sinc, 13–14, 22
weighting, **39**, 41

Gas correlation spectrometer, 142–143
Grating, diffraction, 16, 18–19, 21, 26, 70, 129
Littrow condition, *21*
"Greenhouse" effect, 34, 35, 123
Grounding, importance of, 130
Ground loops, 93
Ground sampling distance (GSD), **26–27**, 80

Humidity, relative, 4, 5
Hydrostatic equilibrium, **33**

Image motion, 28–29, **68**, 126
Impulse response, 20, 41, 137, 142
Index of refraction, 1, 9, 85, 89
Infrared spectrometry, 3, 19–20, 29, 67
Instrumental line shape (ILS), 20, 39, 41, 142
of Fourier transform spectrometer, **22**
of grating spectrometer, **22**
Integral Fourier transform, 10
Interference, 5–7, 24, 68, 91, 93 142
Interferogram, 11, 14, 28, 55, 62, 65, 68–70,
74, 78, 80, 82, 89, 94, 96, 98, 99
Interferometer, Michelson, 6, 7, 9, 15, 23,
55, 80, 83
IRIS FTS, 15, **101–104**, *103–105*

Jacquinot advantage, 19
Jones (unit), 60

Kayser, 4
Kirchhoff's law, 38

Lapse rate, **32**
Laser, 69, 71, 78, 80–81, 83, 93, 108
He/Ne, 23, 69, *72*, *73–74*, *108*, *114*, *118*
Nd:YAG, 69, *121*

Linearity, requirements on, 94, 96–98, *97*
Local thermodynamic equilibrium (LTE),
33, 113
Lorentz broadening, 48, 50

Magnification, 27–28, 80
Mark IV FTS, **114–121**, *119*
Maximum path difference (MPD), **13–14**,
20, 82–83
Mesosphere, 32
Meteorology, 3–4, 107
Michelson interferometer, 6, 7, 9, 15, 23, 55,
80, 83
Micrometer, 4
Micron, 4
Mixed pixel problem, 53
Mixing layer, 31
Mixing ratio, 4, 42
Modeling, 34, **36–44**, **52–54**, 113
Modulation efficiency, 24, 25, 62, 85
Molecule, **44–48**, 102, 116, 142
asymmetric rotor, 46
spherical rotor, 46
symmetric rotor, 46
Molecular anharmonicity, 45
Molecular centrifugal stretching, 45
Molecular line shape, **47–52**, *49*
Molecular line strength (intensity), 46
Molecular P, Q, and R branches, 47
Molecular rotation, 45
Molecular temperature dependence, 31,
45–47, *47*, 48, 51
Molecular vibration, 45
Multichannel advantage, 20
Multiplex advantage, 20
Multiplexing, 19–20
frequency-division, 19–20
time-division, 19

Nanometer, 4
Noise:
background photon shot, *65*, **66–67**, *82*, 87
due to image motion, 28–29, **67–68**, 126
due to impaired modulation, **25**
electrical, 59, 62, 65, 82, 84, 86
signal photon shot, 20, **55–57**, 60, 62, **66–67**

Optical path difference (OPD), 7, 9–14,
    20–23, 69–70, 73, 78, 80–81, 83
  control of 7, 23, 69, 77–78, 82, 84, 93,
    102, 108, 135
  rate of change of (RCOPD), 69, 73
Optical thickness, 89

Parts per billion (ppb), 4, *35*
Parts per million (ppm), 4, *35*
Parts per trillion (ppt), 4, *35*
Parts per million·meters (ppm·m), 4
Path difference, 7, 9–11, 13, 14, 20, 21, 23,
    69, 70, 73, 78, 80, 82, 84, 89, 93, 98,
    104, 116
  zero (ZPD), 14, *74*, 78, 82–83, 89, *90*, 91, 98
  maximum (MPD), **13–14**, 20, 82–83, 98,
    102, 108, 114, 118, 121
Phase, **2–3**, 6, 108
  change, 9, 89
  difference, 6–7, 9
  dispersion, 78, **89–91**
Pixel, 26, 53, *121*
Planck function, 37, 41, 56, 63
Pointing jitter, 28, **68–69**, 107
Postdisperser, cryogenic, 67
Precipitable millimeters, 5
Principle of superposition, 2, 7
Profile:
  compositional, 31, **35–36**, 101, 113, 116
  temperature, **31–33**, 42, 101
"Push-broom" imaging, 29, 69

Radiance, **36–38**, 41, 42, 65, 80
Radiative transfer, 31, **36–44**, 51
  equation of, *39–42*, 55
Radiometric accuracy, 77, 93, *95–97*, 97
Radiometric models, **60–68**, *61, 64–65*
Range, dynamic, *65*, 87, **94**
Reciprocal centimeter (cm^{-1}), 4
Refractive index, 1, 9, 85, 89
Relative humidity, 4, 5
Remote sensing, 3, 4, 22, 23, 24, **31–54**, 55,
    57, 61, 67, 83, 89, 99, 100, **102–127**,
    129–133, 135, 142
Resolution, spatial, 26, 36, 54, 113, *see also*
    Spatial resolution
Resolution, spectral, 9, 13, 17–20, 39, 41,

    51–52, 62, 67, 73, 137, *see also*
    Spectral resolution
Retrieval:
  of atmospheric parameters, 34, **36–44**, 113
  of surface parameters, 39, 42, 44, **52–54**
Retroreflector, **83–84**, 85, 87
  cat's-eye, **84**, *108, 114*
  cube corner, **84**, 121, *121*
Retrosurface, 89

Sampling, **69–83**
  control of, **69**
  errors in, **75–76**, *75*
  interval, *72*
  methodology, **77–79**
  rate, 79, *99*, 108
  strategy, 68, **76–77**
  theory of, 14, **70–75**
Scale height, 33, 36, 40
Scattering:
  aerosol, 37, 44
  instrumental, 58, 60
Sensors, 34, 51, 54, 55, 57, 97, 113, 124, 132, 135
  downlooking (nadir), 31, **36**, *36*, 39, 42,
    73, 101, 123, 132
  limb, 27, **36**, *36*, 42, 80, 113, 126
  occultation, **42**, 97, 113
Shear, beam, **85**, 87
signal-to-noise ratio (SNR), 84, 86, 93–95
  estimation of, 55–68
  impact of image motion on, **68–69**
Sinc and boxcar functions, 11, 13–14, 22
Single-event upset (SEU), 133
Snell's law, 85
Solid angle, 15, **17–19**, 41
Sounding:
  limb, 27, **36**, *36*, 42, 80, 113, 126
  near-nadir, 31, **36**, *36*, 39, 42, 73, 101, 123,
    132
  solar occultation, **42**, 97, 113
Sources:
  broad band, 7, **10**
  calibration, 24, 98
  finite size, **8–9**, 18
  point, 7, 91
  remote sensing, 42, **55–57**, 62–63, 65, 67,
    87, 97

Spacelab 3 ATMOS FTS, 51, 101, **109–114**, 116, *115–116*, 123, 127

Spatial coverage:
imaging FT spectrometer, **27–28**
imaging grating spectrometer, **28–29**

Spatial resolution:
at limb, 36, 113
on ground, 26, 36, 54

Spectral resolution, 9, 13, 17–20, 39, 41, 51–52, 62, 67, 73, 137
Fourier transform spectrometer, **13**, 17–18, 20, 22, 26–28, 52, 67, 73, 81, *99, 102, 108, 114, 118, 121*
grating spectrometer, **18**, 21–22, 67

Spectrometer:
cooled array, 67
dispersive, 15–16, **18–19**, 20–26, 28–29, **66–69**, 93, 129, 142
Fourier transform, 1, 11, 13, 15–16, 19–20, **23–29**, **55–100**, *102, 108, 114, 118, 121*
gas correlation, 142–143
imaging, 13, 22, **25–29**, 79, 109, 121, 124, 131

Spectrometry, infrared, 3, 4, 19, 20, 22, **31–59**

Starting, 28, **69**

Stratosphere, 32, 34, 36, 51–52, 113, 116, 123, 132

Stray light, 57, 80, 74

Superposition principle, 2, 7

Surface (Earth), 27, 31–33, 39, 41–42, 44, 52–54, 57, 65, 100, 113, 123

Telescope
Canada–France–Hawaii, 107
collecting, 80
use of FTS at, 67, 79–80, 104–109, **127–131**

Temperature:
atmospheric, 4–5, 22, **31–34**, *32*, 37, 41–42, 46–48, 51, 101, 123
detector, 58–60, 62–63, 67, 82, *102, 108, 114, 118, 121*
surface, 38, 41–42, **56–57**, 63

system brightness, 58, *58*, 60–61

TES FTS, **121–127**, *122–124*

Thermosphere, 32–33

Thickness, optical, 89

Throughput, 15, 66

Transform, Fourier, 8, **10–14**

Transmittance:
atmospheric, 34, 37–39, 41–42, *43*, 44, 50, 65
instrumental, 15, 24, 60–63, 80, 89, 91

Troposphere, 32, 36, 52, 123

Units, 1, 3, 4, 7, 36, 45, 49, 77
of atmospheric physics and chemistry, **4–5**
of detectivity, 60
of spectrometry, 4

Velocity of light, 1, 23, 45

Vibration, 131
amplitude modulation by, 28, **68–69**, 107
sampling errors induced by, 75–76, *75*, **91–93**

Voigt function, 50 51

Voyager IRIS FTS, 15, **101–104**, *103–105*

Wave equation, 1

Wavefront, **6–7**

Wavefront division, 6

Wavelength, 1, 4, 9, 17, 19–26, 44, 52, 54, 57, 59–60, 67, 71, 78, 83, 108, 123

Wavenumber, 4

Waves:
electromagnetic, 1
superposition of, 2, 5, 7

Weighting function, **39**, 41

"Whisk-broom" imaging, 29, 69

Window, atmospheric, *43*, **44**, 67

Zero path difference (ZPD), 14, 78, 82, 83, 89, 91, 98

(*continued from front*)

Vol. 63. **Applied Electron Spectroscopy for Chemical Analysis.** Edited by Hassan Windawi and Floyd Ho

Vol. 64. **Analytical Aspects of Environmental Chemistry.** Edited by David F. S. Natusch and Philip K. Hopke

Vol. 65. **The Interpretation of Analytical Chemical Data by the Use of Cluster Analysis.** By D. Luc Massart and Leonard Kaufman

Vol. 66. **Solid Phase Biochemistry: Analytical and Synthetic Aspects.** Edited by William H. Scouten

Vol. 67. **An Introduction to Photoelectron Spectroscopy.** By Pradip K. Ghosh

Vol. 68. **Room Temperature Phosphorimetry for Chemical Analysis.** By Tuan Vo-Dinh

Vol. 69. **Potentiometry and Potentiometric Titrations.** By E. P. Serjeant

Vol. 70. **Design and Application of Process Analyzer Systems.** By Paul E. Mix

Vol. 71. **Analysis of Organic and Biological Surfaces.** Edited by Patrick Echlin

Vol. 72. **Small Bore Liquid Chromatography Columns: Their Properties and Uses.** Edited by Raymond P. W. Scott

Vol. 73. **Modern Methods of Particle Size Analysis.** Edited by Howard G. Barth

Vol. 74. **Auger Electron Spectroscopy.** By Michael Thompson, M. D. Baker, Alec Christie, and J. F. Tyson

Vol. 75. **Spot Test Analysis: Clinical, Environmental, Forensic and Geochemical Applications.** By Ervin Jungreis

Vol. 76. **Receptor Modeling in Environmental Chemistry.** By Philip K. Hopke

Vol. 77. **Molecular Luminescence Spectroscopy: Methods and Applications** (*in two parts*). Edited by Stephen G. Schulman

Vol. 78. **Inorganic Chromatographic Analysis.** Edited by John C. MacDonald

Vol. 79. **Analytical Solution Calorimetry.** Edited by J. K. Grime

Vol. 80. **Selected Methods of Trace Metal Analysis: Biological and Environmental Samples.** By Jon C. VanLoon

Vol. 81. **The Analysis of Extraterrestrial Materials.** By Isidore Adler

Vol. 82. **Chemometrics.** By Muhammad A. Sharaf, Deborah L. Illman, and Bruce R. Kowalski

Vol. 83. **Fourier Transform Infrared Spectrometry.** By Peter R. Griffiths and James A. de Haseth

Vol. 84. **Trace Analysis: Spectroscopic Methods for Molecules.** Edited by Gary Christian and James B. Callis

Vol. 85. **Ultratrace Analysis of Pharmaceuticals and Other Compounds of Interest.** Edited by S. Ahuja

Vol. 86. **Secondary Ion Mass Spectrometry: Basic Concepts, Instrumental Aspects, Applications and Trends.** By A. Benninghoven, F. G. Rüdenauer, and H. W. Werner

Vol. 87. **Analytical Applications of Lasers.** Edited by Edward H. Piepmeier

Vol. 88. **Applied Geochemical Analysis.** by C. O. Ingamells and F. F. Pitard

Vol. 89. **Detectors for Liquid Chromatography.** Edited by Edward S. Yeung

Vol. 90. **Inductively Coupled Plasma Emission Spectroscopy: Part I: Methodology, Instrumentation, and Performance; Part II: Applications and Fundamentals.** Edited by J. M. Boumans

Vol. 91. **Applications of New Mass Spectrometry Techniques in Pesticide Chemistry.** Edited by Joseph Rosen